EX LIBRIS
SAGAX·SCRUTANDO·SAGACIOR
BRYAN L. SAGE

THE

CENTRE

The Natural History of Australia's Desert Regions

An emu footprint in cracked clay cradles saltbush seed-bearing bladders.

Penny van Oosterzee

Reg Morrison

REED

To the memory of those
desert dwellers which
disappeared
even before we knew
they were there.

First published in 1991 by
Reed Books Pty Ltd
3/470 Sydney Road Balgowlah
NSW 2093

National Library of Australia
Cataloguing-in-Publication Data

van Oosterzee, Penny
The Centre, a natural history of
Australia's desert regions

Includes index
ISBN 0 7301 0309 9

1. Natural History — Australia, Central
I. Morrison, Reg. II Title.
508.942

Produced in Australia by the Publisher

Edited by Arlené Santoso

Designed by Lawrence Hanley

Proof-reading by Dawn Hope

Illustrations by Alistair Barnard

Typeset in New Zealand by Deadline

Printed in Singapore through
Imago Productions

ACKNOWLEDGMENTS

There are two cherished friends and colleagues to whom I am completely indebted. They are Peter Mitchell and Steve Morton. They have helped me in so many different ways. Peter laboured through the draft and helped with diagrams. Steve provided many of the references and checked text. Both lovers of Australia's arid zone, they offered unstinting support and advice.

Australia has many other excellent and enthusiastic scientists who are making pioneering attempts to understand the many fascinating aspects of Australia's heartland. A substantial number of these people helped me understand the often revolutionary concepts underpinning the geology, natural history and pre-history of the arid zone. Some of them may not even remember the particular phone call or reference which, for me, threw open yet another blind. People who help in projects such as these deserve to be acknowledged and I'm deeply sorry if I've missed anyone. Thank you Xiang Yang Chen, Bob Chinock, Peter Copley, Dick Evans, Dean Graetz, Russell Grant, John Greenslade, Graham Griffin, Trevor Hobbs, Jeanette Hope, Surrey Jacobs, Ken Johnson, Geoff Lundie Jenkins, Peter Murray, Julian Reid, Gurdip Singh, Mike Smith, Mark Stafford Smith, Bruce Thomson, Elizabeth Truswell, Bob Wasson, William D. Williams, Stephen van Leeuwen.

My warmest thanks to the team at Reed Books. A more helpful, friendly and supportive group you couldn't wish for. Thank you, too, to Lawrence Hanley for making the book beautiful.

Noel Preece, my beloved husband, always seems to come toward the end in these things, which is entirely unfair. Yet I deliberately did not let Noel help with the book because I knew that if he did he would never read the final product. So now, Noel, you have no excuse! And finally to my beautiful boy Luke for his exuberant love of life and pride in his Mum.

One of my most memorable impressions of central Australia was from the air, on my first flight to Alice Springs. I'd been watching the subtly patterned, apparently waterless grey-green landscape. Suddenly, out of the heat haze from the north-west, coiled a broad, shimmering river, not of water, but of sand. To me, it was the embodiment of the mystery and magic that is central Australia. In 1988, the same river of sand, the Finke, was in flood — as the locals put it, it had 'come down'. It was the sixth time in a century that you could have seen the entire length of the river in flood — from the headwaters (opposite, top left) to the Simpson Desert. The other photographs show the middle reaches of the Finke centred around the painted cliffs of Horseshoe Bend. The cliffs are crowned by dunes blown from the Simpson Desert.

INTRODUCTION

Arid Australia is 70% of the continent. It is synonymous with the outback, the ethos of all Australians. Yet there has never been a natural history written about it.

This is mainly because there is so little information about Australia's deserts — that part of the country that cannot support crops and has an average rainfall of less than 250 mm. To give an example, for regions such as the central ranges there is still not a better text than *The Horn Scientific Expedition to Central Australia* written by the polymath Baldwin Spencer in 1896. For the saltbush and bluebush plains and the tussock grasslands, Australia's sheep and cattle rangelands, virtually no information exists on the wildlife. By the time we did decide to take a look, many native animals and most of the medium-sized mammals had vanished. In fact Australia is in the invidious position of having lost more mammal species than anywhere else on earth. Most of these are desert dwellers which vanished behind our backs.

For too long the peculiarities of our arid zone have been clouded by comparisons with other world deserts, particularly North America. Compared to the badlands of North America, for instance, arid Australia has a depauperate rodent fauna (which was not always the case). Its reptile radiation, carnivorous marsupials, climatic unpredictability and deceptive flatness were confusing. It was considered a strange dry land.

In the past few years, however, our own scientists have been generating an ecological revolution. Now a whole new set of home-bred ideas is challenging the way people think about the Australian deserts. Many of the concepts presented in this book stem from a handful of dedicated scientists who work in the centre of the continent. The 'dead heart' is a myth. To the contrary, Australia's deserts are rich and varied, as this book reveals.

The flow of water from the desert mountains to the great sump of Lake Eyre provides the conceptual thread for this book. Part One uses one of the oldest rivers of the world, the Finke River, to introduce the arid zone. Using the river as a microcosm of the arid zone we trace the geological history of central Australia and introduce the current ecosystems which comprise Australia's deserts. Part Two explores each of these ecosystems and their intriguing wildlife, from desert mountains to salt lakes.

Rainfall is the unpredictable maestro of Australia's desert regions, driving both the physical and biological forces of the inland.

CONTENTS

PART ONE

PART TWO

ECOSYSTEMS OF THE INLAND

Landmarks of the Inland

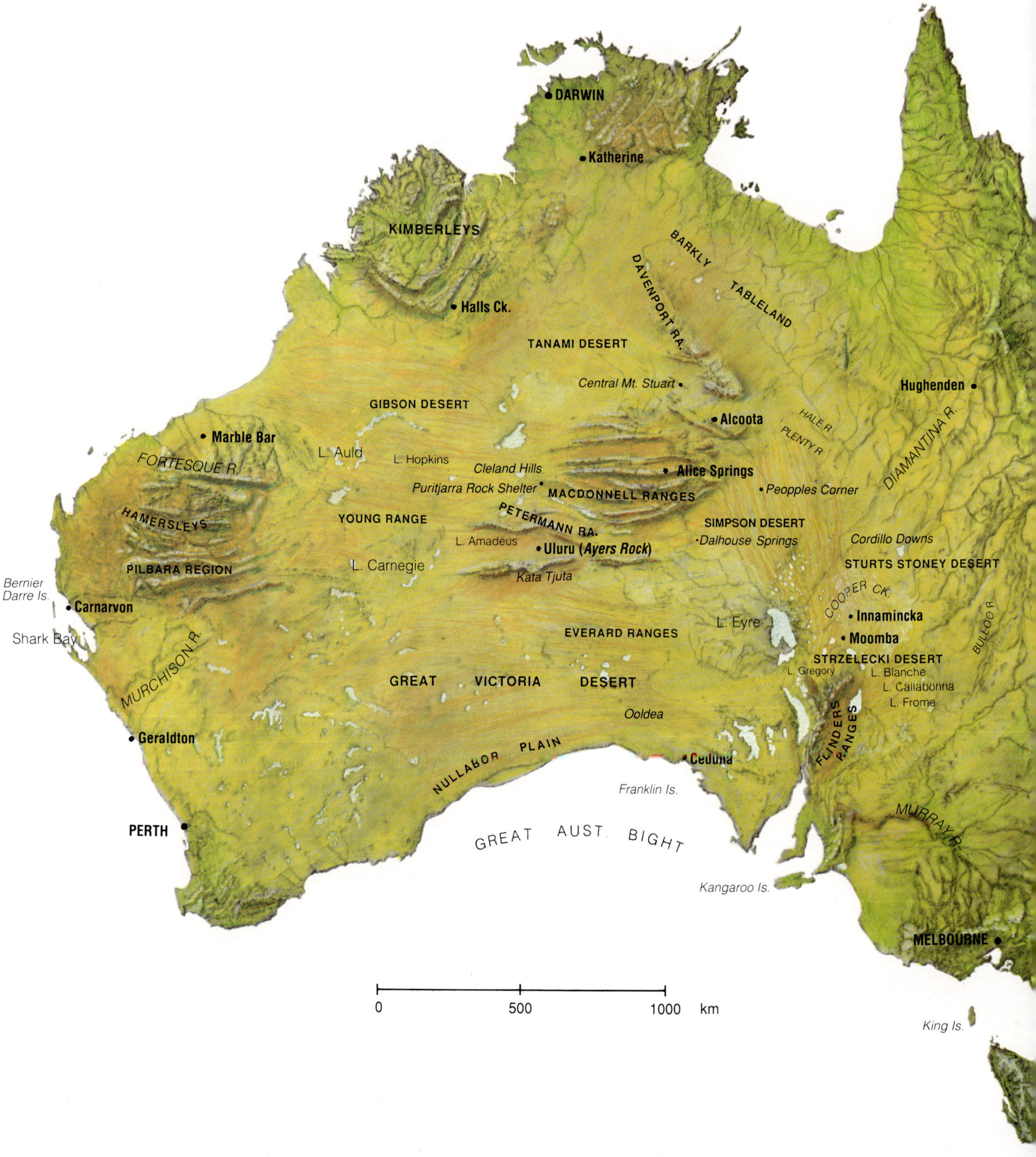

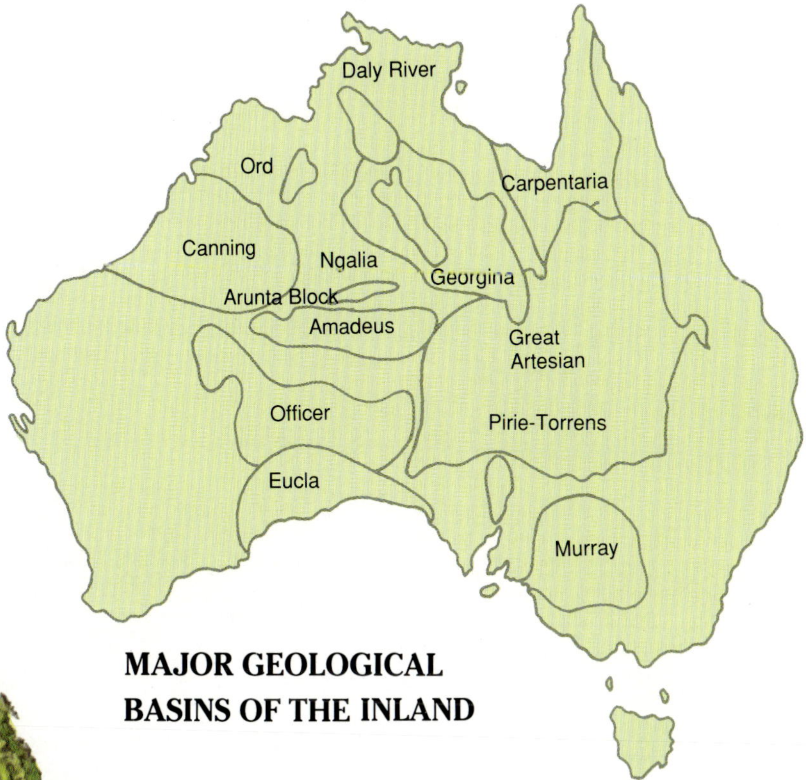

MAJOR GEOLOGICAL BASINS OF THE INLAND

COONGIE LAKES SYSTEM

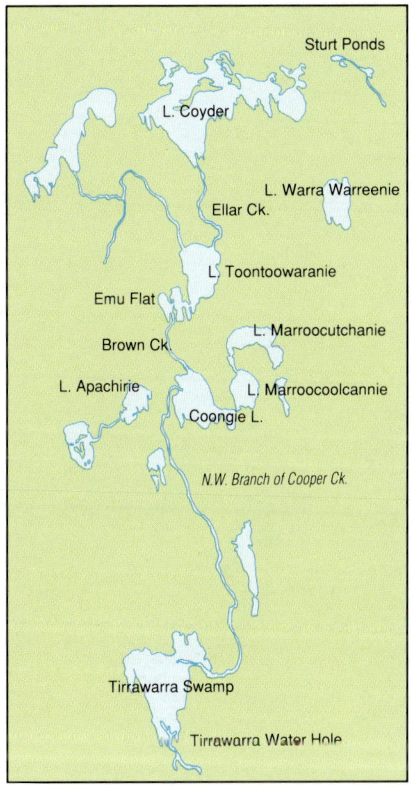

CENTRAL AUSTRALIA

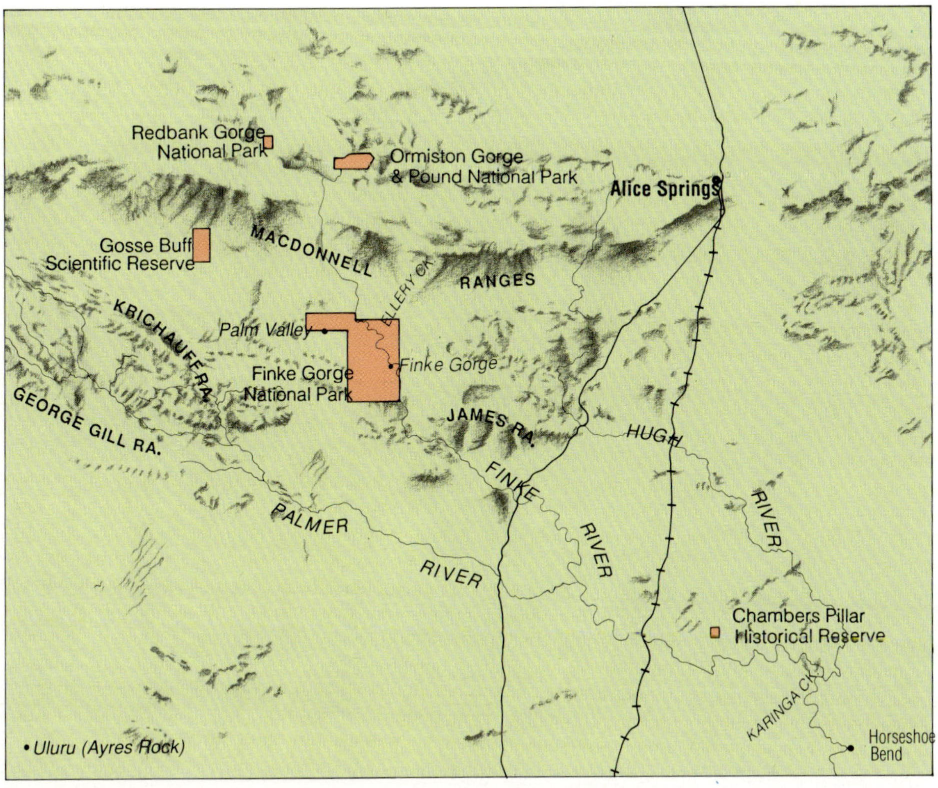

Landsat imagery by kind permission of the Australian Centre for Remote Sensing, Australian Surveying and Land Information Group, Department of Administration Services, Canberra.

Part One

A Strange Dry Land

No river, but the trace

as of dried tears

on a worn face

Frederick T. Macartney,
'Desert Claypan'

This is the driest part of the driest habitable continent. Yet it is still traced by (mostly dry) rivers. The Warburton River slices diagonally across this satellite image of the Simpson Desert and Sturt's Stony Desert. The enormous parallel sand dunes of the Simpson Desert run north-west. Extensive areas of salt lake outline ancient shorelines of a once vast freshwater lake system. The higher gibber-covered ground of Sturt's Stony Desert appears as brown-black.

Nude, smooth, and giant-huge,
the torsos of the gums
hold up the vast dark cave
as the great moon comes.

Shock-headed black-boy stands,
with rigid, thrusting spear,
defiant and grotesque
against that glistening sphere.

In clenched, contorted birth
black banksias agonise;
out of the ferns and earth,
half-formed, beast-boulders rise;

because The Bush goes back,
back to a time unknown:
chaos that had not word,
nor image carved on stone.

Roland Robinson, 'Altjeringa'

The Long Grand Canyon

Things are simply different in arid Australia: it is a land driven by extremes rather than averages; a flat land which is richly patterned; an infertile land which has produced a wealth of plants and animals. It is a land inhabited by lizards rather than mammals.

No other desert system experiences the same climatic capriciousness or the exaggerated impact of the sterile soils. What has survived the ravages of time is a 'perverse' assemblage of plants and animals.

In this vast, timeless land the very grains of the continent have been worked by wind and water over eons. In their antiquity the soils are unique in the world; no other country has quite the array of deficiencies of trace elements in such a diversity of soils.

The weather-beaten soils reflect the ancient rocks of the continent. In Australia, for instance, we have a fragment of the oldest known rocks in the world. These 4000 million year old rocks, found in Western Australia, are remnants of the world's primeval crust, the first scum on the upwelling plumes of molten material which would eventually coalesce to form 'shields'; the cores of the infant continent.

The western two-thirds of Australia is composed of these Archaean to Middle Proterozoic shields. The shields are gently contoured, in places forming massive basins. Like gargantuan soup bowls these basins have been filled with sediment mostly laid down in shallow seas which more or less covered the continent for much of geological time. The Canning and Officer basins are examples of such bowls of sediment and, today, contain the Great Sandy Desert and the Great Victorian Desert.

The Amadeus Basin in the very centre of the continent, cradling Uluru (Ayers Rock) and Kata Tjuta (The Olgas), was also formed at this time. The basin is slashed by one of the oldest rivers in the world, the Finke River (or Larapinta to the Arrernte people). And here, miraculously, is a window to view the geological history of arid Australia and to see many of the processes which have moulded the arid zone. Ellery Creek, a tributary of the Finke, reveals, like an open book, an 1800 million year old slice of central Australia's history.

By incredibly good fortune, in a section only ten kilometres long, the sequence of rocks through which the Ellery cuts has been tilted through 90 degrees; rather like a wall of the Grand Canyon turned on its side. So instead of taking a donkey ride down the Grand Canyon, a vertical mile of layered rock which from top to bottom represents 2000 million years of Earth's history, it is possible to jump into a four-wheel drive and drive along Ellery Creek.

***Kata Tjuta and, in the background, Uluru are carved from debris washed off the once stupendous Petermann Ranges around 600 million years ago. Above right, the fantastic-looking Thorny Devil* (Moloch horridus) *and, right, the Frilled Lizard* (Chlamydosaurus kingii).**

Red grains of sand have been stripped from the MacDonnell Ranges and carried downstream by the Finke River. Here, they have been piled into dunes on the north-west extremity of the Simpson Desert at Horseshoe Bend.

The first runnels of Ellery Creek trickle across the ancient central Australian shield, the Arunta Block. Sometime before 1800 million years ago, these rocks were originally blasted out of volcanoes and then swept by proto-oceans which left a veneer of fine-grained sediments. The rocks have undergone at least two cycles of compression, at 1800 million years ago and 1600–1700 million years ago, when they were squeezed like geological toothpaste and buried. The two mountain ranges that these events represent had come and gone before the (already ancient) shield sank beneath an extensive sea covering a shallow marine shelf 900 million years ago.

A blanket of coarse sand 800 metres thick, made up in part from the residues of these ancient mountains, was deposited on top of their worn down roots. The sand was later buried and heated — virtually pressure cooked — before being exposed once again as a quartzite known as Heavitree Quartzite. This section of rock forms the orange cliffs shadowing Ellery Bighole, just south of the Arunta Block. The ripple marks of 900 million year old gently lapping ocean waves can still be seen in the spectacular cliffs of this quartzite.

Just south of Ellery Bighole is a carpark. It has been graded on the sediments of what used to be a shallow lake choked with mats of cyanobacteria (a bacteria which can photosynthesise and make its own food). Very near the carpark is a ledge exposing remnants of these lakes: slabs of fine-grained rock, subtly coloured greys and pinks. They tell a story of hot, cloudless skies and clear still water . . .

The mounting blanket of wave-deposited sand, which was to become Heavitree Quartzite, was slowly blocking the seaways. Landlocked lakes and basins developed on top of the sand, scattered under a hot moisture-sucking sun. Evaporation of these bodies of sea water resulted in the water becoming super-saline. Layers of different minerals crystallised at the bottom of the basins. Episodes of storm and rain swept in mud and fine debris from the surrounding undulating land. The thin slabs of blue-grey shale and the thicker slabs of pink siltstone (all part of what is known as the Bitter Springs Formation) in the ledge near Ellery Creek carpark are the visible evidence of these events.

Noon sunrays set nearly billion-year-old Heavitree Quartzite aglow with colour at Redbank Gorge in the MacDonnell Ranges.

Slabs of fine-grained limestone tell a 900 million-year-old story of hot, cloudless skies and clear, still water. Much later, the contortions which resulted in the MacDonnell Ranges threw these horizontal slabs of rock upright.

Above, six hundred million years ago, in the sea through which filtered the sand that would eventually become Uluru, lived jellyfish like the ones found here at Ediacara in the Flinders Ranges.

Left, columns of stromatolites were the first forms of life to be fossilised. Nine hundred million years ago, these mats of algae-like material thrived in the seaway that spanned central Australia.

Further down the sandy bed of Ellery Creek, underneath the magnificent shady river red gums, are traces of life which flourished in the ancient super-saline lakes. Light coloured rocks — marbled limestone shot with purple — emerge from the sandy river bed. A closer look will show that the curved purple lines take the form of columns. These are fossilised domes of stromatolites, formed by layer upon layer of matted cyanobacteria.

The time of calm seas and placid lakes came to a sudden (geologically speaking) halt at around 700 million years ago. The climate cooled and the whole of Australia, which had drifted across the globe into polar climates (possibly the North Pole), was in the grips of a glaciation. At the same time there was another uplift in topography. Parts of central Australia became ice-sculpted with knife sharp divides, jagged and serrated crests and large U-shaped valleys. The downhill-sliding ice was peppered with rock particles which had already been through many cycles of weathering — granite from the Arunta Shield, quartzite from the Heavitree Quartzite and limestones from the previously formed Bitter Springs Formation. The glaciers slid into a shallow sea where the rocks, torn from the walls and floors of the valleys and often ground into rock-flour, dropped out and sank to the sea bottom. The continuous rain of unsorted material finally lifted the surface of deposition above sea level. This moraine became an unco-ordinated maze of hills, lakes and ridges washed by melt-waters. The remnants of this period are piled up against the limestone of the Bitter Springs Formation.

Further afield the impressive Mount Conner table-top, which can be seen today along Lasseters Highway on the way to Uluru, is also a remnant of these times.

The climate began to improve as Australia drifted into lower latitudes. Rivers spilled pebbles and sand into massive deltas, and 2000 metres of silt and mud accumulated in the shallow seas still covering much of the continent. Amazingly, individual tides of this period have been trapped in sandstone in the bed of Ellery Creek near the crumbly moraine. The 2000 metres of the silt and mud has since turned to the shale which is exposed in the creek bed near the tide-marked sandstone.

Not a particularly good area for fishing at the best of times, central Australia nevertheless has a rich fossil fish record from the distant past. These strange armoured fishes, called Bothriolepis *come from 360 million-year-old rocks in the western MacDonnell Ranges of central Australia.*

ALEX RITCHIE

Underside of Arandaspis *headshield.*

Topside of Arandaspis *headshield.*

These well-preserved impressions of two segmented flatworms, called Dickinsonia*, are examples of marine animals that lived in seas covering central Australia 600 million years ago.*

THE OLDEST FISH *ARANDASPIS*

In 1959 a scientist poking about on a sandstone ridge on the edge of the Simpson Desert in central Australia discovered an imprint, 1 cm sq, in a sandstone block: part of the headshield of a very ancient fish. From later, more complete, finds an Australian Museum scientist reconstructed and described *Arandaspis* (named after the Aranda tribe of central Australia and 'aspis', Greek for shield). *Arandapsis*, the first discovery of its kind in the Southern Hemisphere, is the oldest known fish in the world.

In Ordovician times, 470 million years ago, *Arandaspis* lived near the shores of a shallow sea which covered much of central Australia. This jawless fish, 12-14 cm long, lacked paired fins and was probably a poor swimmer. It lived on the bottom, swam in short bursts and fed on micro-organisms and organic debris carried by currents just above the bottom sediments. Fossils of related forms have since been discovered in slightly younger rocks in Bolivia, South America

Dawn on Mt Sonder in the MacDonnell Ranges.

Impressions of the gentle waves of an ancient sea, miraculously trapped in rock in the heart of Australia.

THE OLDEST SURFACES ON EARTH

Inland Australia is well known for the flatness of its landscapes and the antiquity of its landforms. And in the Davenport Ranges about 400 kilometres north of Alice Springs it appears that the very surface of some of the land is the oldest on Earth to be persistently exposed to our atmosphere. Five hundred million years ago sand was stripped from ancient sandstone ridges and deposited in valleys that flowed into the sea which surrounded the Davenports at the time. Subsequent dissection has carved up this surface but remnant terraces and mesas have never been covered since.

The weathered granite comprising the Devil's Marbles, 400 km north of Alice Springs, is part of the Davenport Ranges. The flat surfaces of these ranges — one can be seen in the background — are thought to be the oldest surfaces on Earth.

Like a family photo album, this geological timescale marks events which were important in the development of arid Australia, such as its birth, development, global driftings, union with other continents, and the final parting which led, eventually, to the outback of today.

From top left to bottom right, the world's oldest land crust (3.6 billion years), found at Mt Narryer, Western Australia; 2.8 billion year old stromatolites; 750 million year old glacial debris in the Indulkana Ranges; 500 million year old Kata Tjuta; a 500-600 million year old trilobite; 3500 year old Australian Canis familiaris dingo.

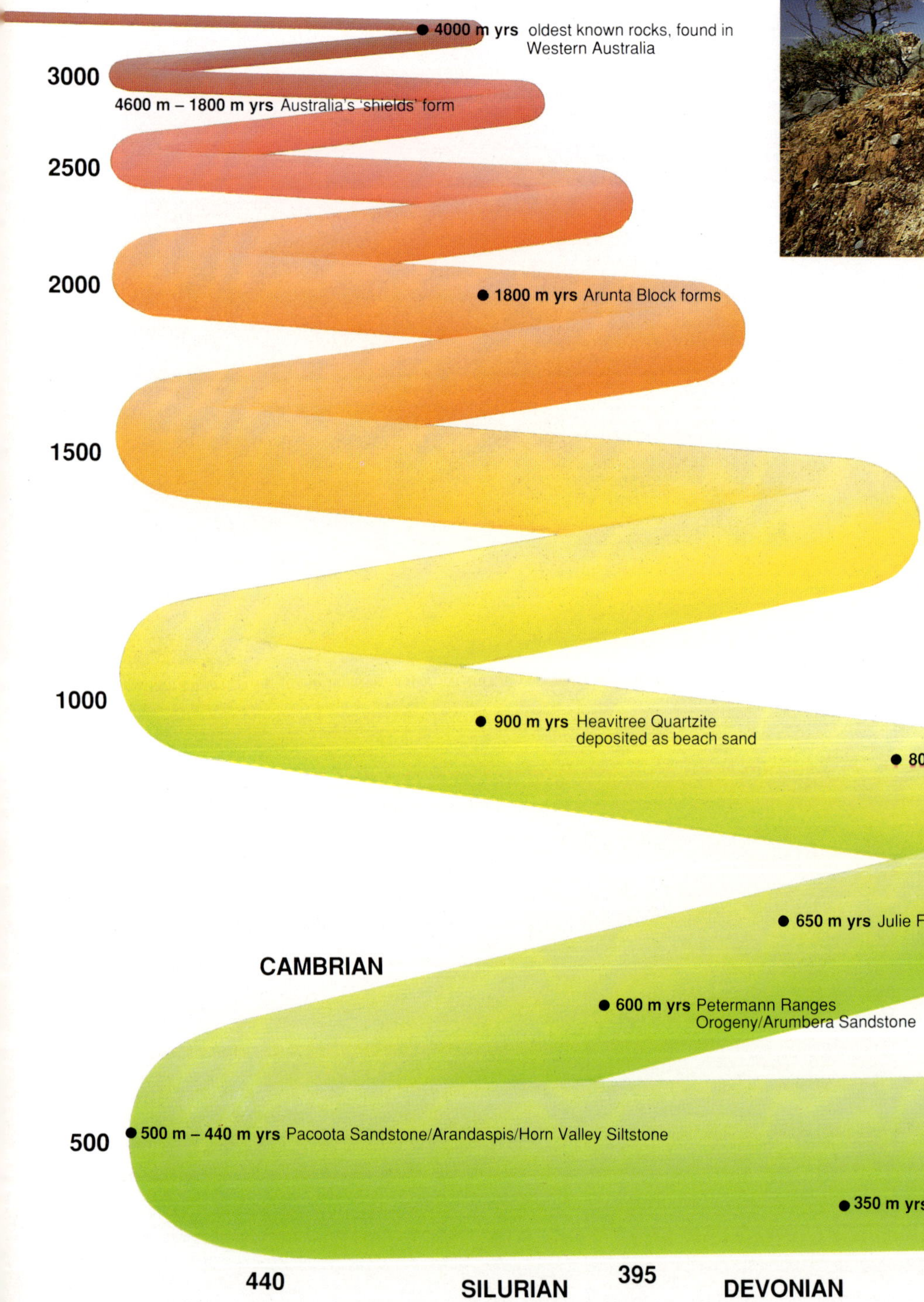

Towards the end of this period the shallow marine shelf was practically filled with sediment. The water became so concentrated that another limestone was precipitated and deposited. This limestone, the Julie Formation, exists as a thin and striking dark-grey ridge up to four metres high. The placid accumulation of limestone-forming ocean ooze was to be violently interrupted with a series of events, the aftermath of which gave birth to one of the world's great landmarks: Uluru (Ayers Rock) and Kata Tjuta (The Olgas).

If an observer could fix a camera to film the Earth's surface by snapping one frame every five hundred years, a spectacular one-hour feature film could be created. This film would condense around 50 million years of geological time and would record a continental collision which would eventually give birth to Uluru and Kata Tjuta. The event is called The Petermann Ranges Orogeny.

Around 600 million years ago a collision, rivalling the Himalayan-collision of India with Asia, pushed up thousands of metres of rock somewhere near the border of the Northern Territory and South Australia. The film would show a ragged mountain range being literally squeezed out of the ground, with magma oozing like blood at the roots of the mountains.

Torrential rivers would appear streaming off the mountains carrying boulders, pebbles and sand and dumping them in a vast alluvial fan on the mountain's northern flanks. These were the raw materials that would, after burial and heating, turn to rock, and millennia later reappear as Uluru and Kata Tjuta.

The one-hour feature film would not only show mountains being born but would also show an ebbing sea regressing from the flanks of the new mountains. A delta would be formed, with the sea reduced to a shallow embayment washing over the area in which Ellery Creek now flows. After debouching from the Petermann Ranges the rivers would wander rather sluggishly through the delta and finally empty their sediment into the sea. This sediment has been transformed into the rich red-coloured ridge of Arumbera Sandstone, just south of the Julie Formation limestone.

Australia lay astride the equator at this time and the shallow, placid sea through which the Arumbera sand filtered was rich in life forms such as jellyfish, burrowing worms and the first brachiopods, or lamp shells. Early ancestors of oysters and mussels, both bivalves, sat rooted in the mud which we see today as a sequence of dark shale just south of the 600 million year old Arumbera Sandstone.

The next rock sequence is again a sandstone interbedded with some limestone, which indicates that the sea was once again in a cycle of transgression. A new animal can be found as a scarce fossil in this outcrop. Trilobites, beetle-like animals whose bodies were composed of three lobes, hence

TERTIARY PERIOD

38 m yrs Ice forms in Antarctica
20 m yrs Diverse Australian fauna
14 m yrs Rapid expansion of glaciers in Antarctica/ possible beginning of present ice-age vast rivers in Australia dry up
8 m yrs Fauna assemblage reflecting climatic change
2.37 m yrs – present Winter rain develops/extreme development of glacial-interglacial cycles/increasing aridity

QUATERNARY PERIOD

2 m yrs Mega-fauna
1 m yrs – present 8 glacial-interglacial cycles
50 000 – 30 000 yrs Last time lakes full in central Australia
22 000 yrs Humans occupying arid zone
18 000 yrs Peak of last glacial/Simpson Desert forms
3500 yrs Dingo in Australia

THE GEOLOGICAL TIMESCALE

- **250 m yrs** Australia, drifting into polar climates, is glaciated
- **130 m yrs** Marine transgression, Great Artesian Basin forms
- **120 m yrs** First flowering plants (angiosperms)
- **65 m yrs** Gondwana break-up
- **50 m yrs** Antarctic Circumpolar Current established/Transantarctic mountains form/extensive waterlogged areas and rivers in Australia

QUATERNARY

AN 225 TRIASSIC 195 JURASSIC 136 CRETACEOUS 65 TERTIARY 2

tri-lobe-ite, swam in the sea or bulldozed their way through the mud on the sea floor. Already there may have been 3000 different species of trilobite from plankton size to seventy centimetres long.

They had well developed heads and were the first creatures on Earth to have high-definition eyes. Their whole bodies were covered in a shield and it is these that are likely to be found embedded in the limestone. Ellery Creek loops through this ridge before heading south through a spectacular sandstone gorge.

The sea was no longer placid. Waves plunged unrelentingly onto the continental shelf. At the same time, movement of the sea floor riding on a swell of magma was causing repeated transgressions and regressions. A succession of longshore bars folded into each other. Along with the coarse sand derived from the breaking waves, these longshore bars resulted in over 1000 metres of sand being piled up. With the sea came legions of bottom-dwelling organisms.

This sand today forms a high escarpment of orange-yellow sandstone, called Pacoota Sandstone. It is riddled with world-class traces of the primeval past. Large sections of rocks, called pipe-rock, are fluted with vertical and U-shaped 'worm' tubes. The scratchings and scurryings of the hordes of trilobites, and even their faecal pellets have been preserved in this rock as coprolitic curiosities. In places, the rock is also corrugated by the ripple marks of 500 million year old waves.

Swarms of animals floated at the whim of wind and currents in these Ordovician oceans. The most numerous of these were the graptolites. Many of them, as they died, sank to be eaten by the army of animals creeping on the bottom. Only a few sank in the soft muds to become fossilised. Graptolite fossils are unpretentious, occurring only as faint impressions amongst the extraordinarily rich variety of other species in the siltstone adjacent to the Pacoota Sandstone. These animals were preserved in the fine oxygen-deficient muds covering the bottom of the deepest parts of the oceans. In these foetid conditions very little decay occurred. As the dead animals rained into the mud it covered them like a shroud and they became remarkably well preserved in the Horn Valley Siltstone.

The sea levels changed once again and pearly white beaches appeared to gleam in a tropical sun. Anyone able to wade in the warm waters would have seen the glimmerings of a biological revolution: the very first fish. The ridge of pale sandstone next to the Horn Valley Siltstone contains rare fossils of this, the first known vertebrate to have evolved.

The adjacent Mereenie Sandstone is one of the most extensive units in the Amadeus Basin. It was deposited, over perhaps 100 million years, in a variety of environments ranging from marine to rivers and lakes, and finally even sand-swept desert dunes. The land was entirely flat — probably a large continental desert blowing sand onto the shores of a shallow sea which slowly crept across the land from the west. Sections of these shores may have been tinged green as early waves of land dwellers, scraps of lichen-like plants, clung

Above, the many heads of Kata Tjuta rise abruptly from the desert plains like islands from the sea.

THE AGE OF ULURU

Uluru (Ayers Rock) and Kata Tjuta (the Olgas) rise abruptly from the surrounding plains like islands from the sea; which is a fitting description for these 'inselbergs', a German word meaning island mountains. How they came to be inselbergs, however, has always been a mystery. For a start they are made out of sedimentary rock which is unusual because most inselbergs are granite. The sediment from which Uluru and Kata Tjuta are made was deposited as an alluvial fan on the flanks of the infant Petermann Ranges about 600 million years ago. This material was pressure cooked to form sedimentary rock. A prolonged phase of erosion about 200 million years ago brought the inselbergs into relief. They appear to have been exposed gradually as a result of the erosional lowering of the surrounding plains — they are, in fact, still growing. While they owe their survival to their massive structure and the absence of open joints, why only they survived while a great thickness of rock disintegrated from the adjacent area is still unclear.

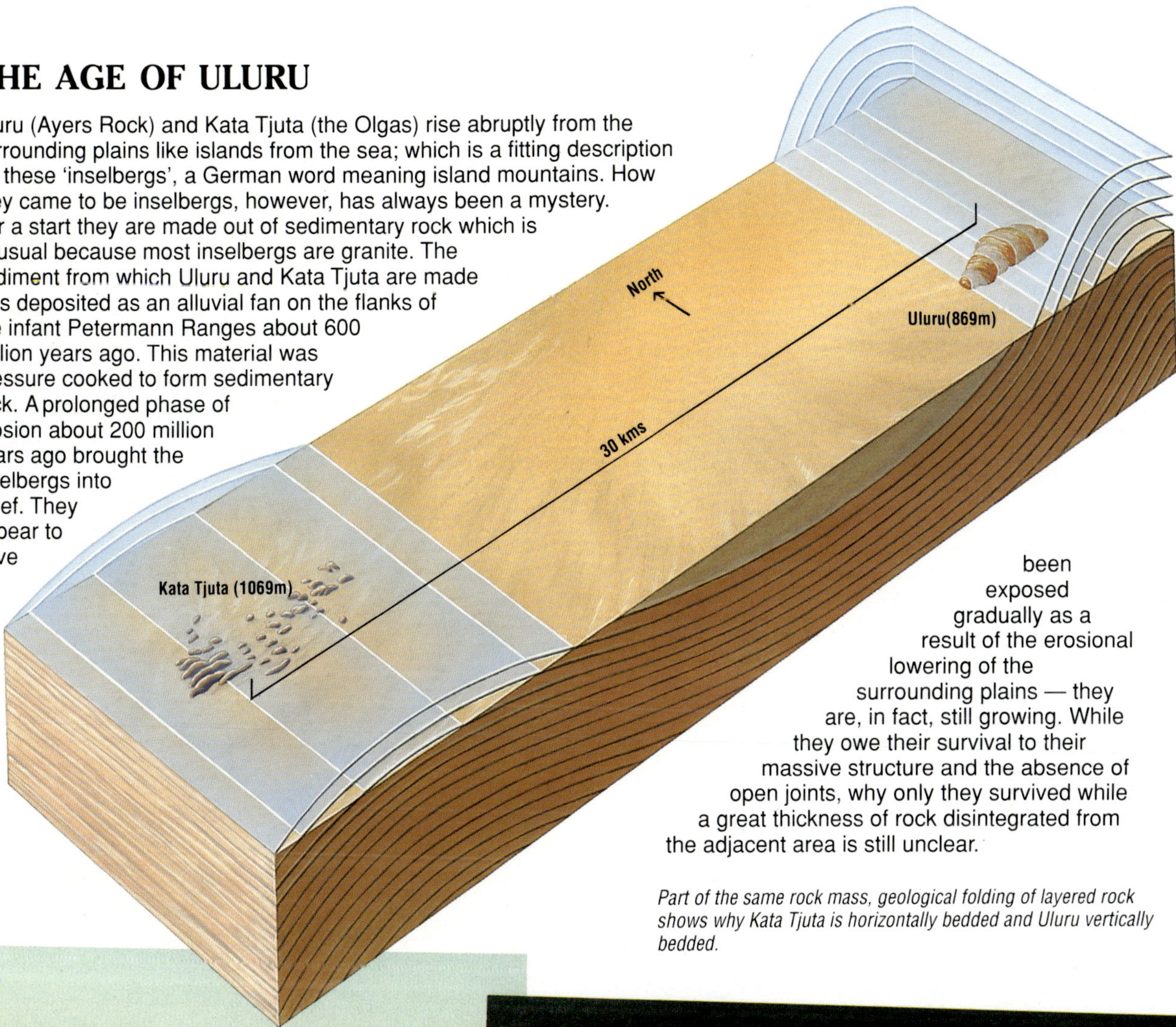

Part of the same rock mass, geological folding of layered rock shows why Kata Tjuta is horizontally bedded and Uluru vertically bedded.

L. HANLEY

Exposed gradually as a result of erosional lowering of the surrounding plains, Uluru is still growing.

tenaciously to the infertile sands.

It is interesting to point out that it was only at this time that the entire eastern part of Australia, a region of volcanic arcs swept by oceans, entered a mountain-building phase which would convert it to stable continental crust. This event may even have been related to the birth of the MacDonnell Ranges in central Australia.

Why a high mountain range (the MacDonnells were at least as high as the Canadian Rockies) should have formed in the centre of a stable continent at this time is a riddle. One explanation involves a vast block of crust from the Kimberleys to the Simpson Desert rotating around a pivot point in the Tanami Desert (about midway). One end of the crust opened, forming the Kimberley trough and the other end correspondingly squeezed closed, forming the MacDonnells.

The final three rock formations, a siltstone, a sandstone and a massive layer of cemented pebbles and boulders, through which Ellery Creek loops, mark this event and its aftermath. Combined, these sediments are about six kilometres thick and comprise the materials that were subsequently torn from the mountains. The cemented pebbles and boulders, today known as the Brewer Conglomerate, formed a distinct alluvial fan on the southern flanks of the MacDonnell Ranges and the rock particles they contain are a mixture of all the formations that the Ellery cuts.

With its load of gravel and sand Ellery Creek joins the Finke within the Krichauff/James Range complex. These latter two ranges are giant folds radiating like tidal waves south of the MacDonnells after its emergence. Together these three ranges comprise the central ranges complex. Intriguingly, the Finke slices through these east–west trending ranges as if they did not exist, looping and twisting to the south. And indeed the Finke had probably established its present course before the Krichauff and James ranges emerged, the river erosion easily keeping pace with the folds rising around it. The banks of the infant Finke River may have been patterned with footprints of the first animals to emerge from the water: primitive millipedes, spiders, centipedes and scorpions.

Two more important episodes can be added to this extraordinary sketch of history profiled by Ellery Creek. One was a glaciation during the Permian, when Australia had drifted to the South Pole. This resulted in a further cycle of debris being left on the landscape to be redistributed by wind and water. The other episode, later, was a marine transgression during the Cretaceous which swamped a major portion of the continent outlined today by the Great Artesian Basin and incorporating Lake Eyre. The depression finally became choked with sediments that pushed the sea from the continent, never to return. The final muds and silts, underlying the Channel Country and the Lake Eyre Basin, were deposited by sluggish rivers.

Apart from these localised fertile muds and silts, inland Australia was mantled by poor, sandy soils; soils that were, as far as nutrients went, exhausted.

Above, still impressive, today's MacDonnell Ranges are the mere worn-down stubs of once Canadian Rockie-sized mountains. Left, these low hills, like a crumpled skirt flanking the MacDonnell Ranges, are in fact debris clawed from the once massive ranges and dumped at their feet.

FORMATION OF MACDONNELL RANGES

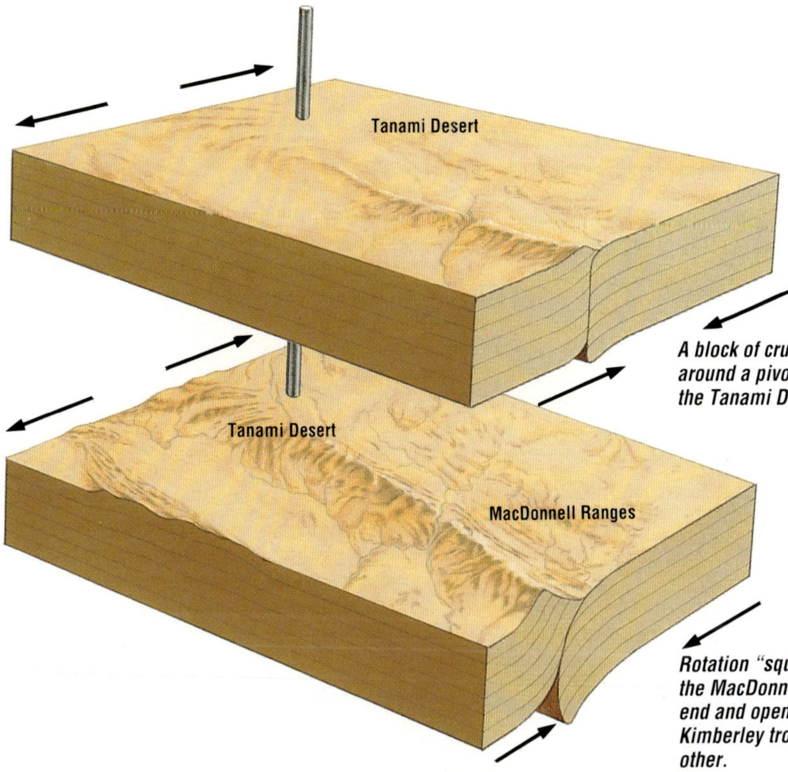

A block of crust rotates around a pivot point in the Tanami Desert.

Rotation "squeezes" out the MacDonnells at one end and opens the Kimberley trough at the other.

THE LONG GRAND CANYON

Like the Grand Canyon turned on its side, Ellery Creek exposes a 2000 million year old chunk of Earth's history in a ten kilometre slice through the MacDonnell Ranges.

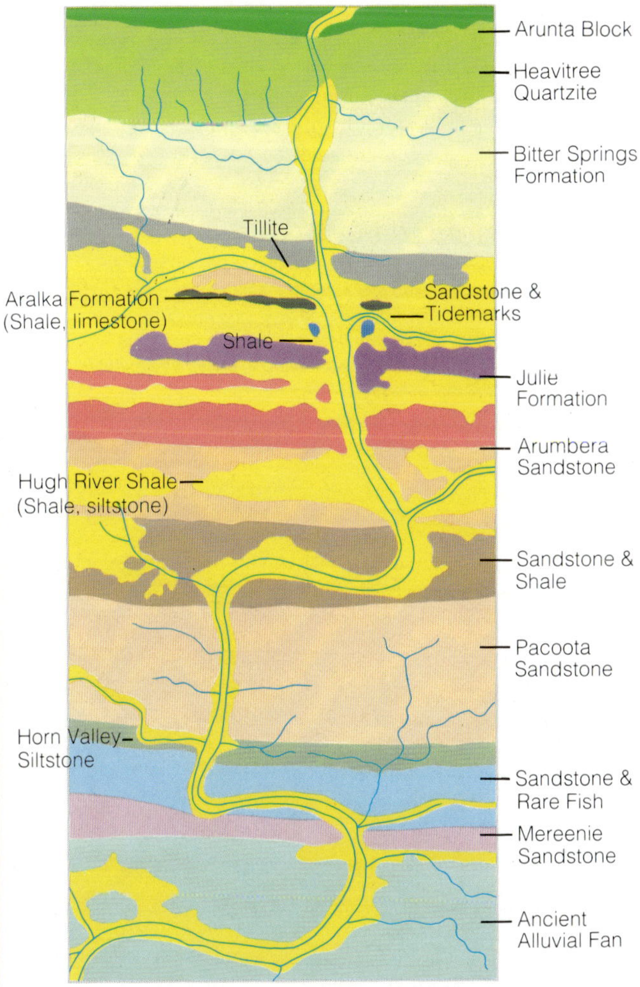

FINKE RIVER

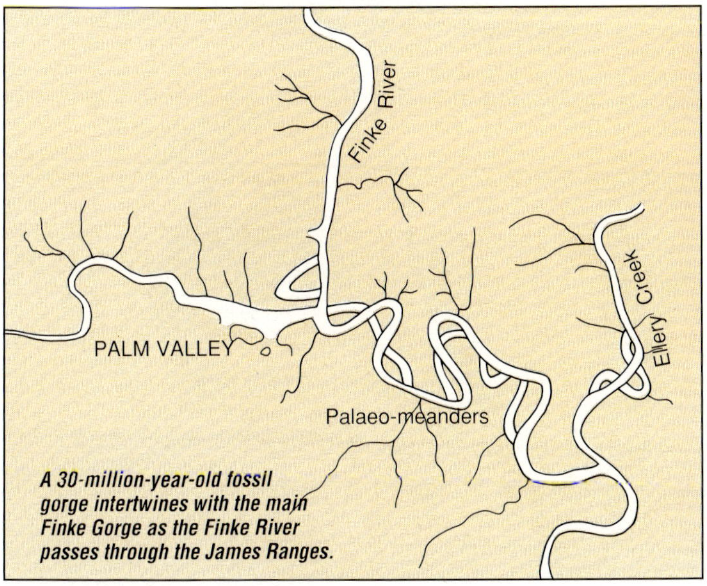

A 30-million-year-old fossil gorge intertwines with the main Finke Gorge as the Finke River passes through the James Ranges.

Evolution of an Enigma

At the beginning of the Tertiary period, 65 million years ago, Australia as part of Gondwana lay near the South Pole. The surrounding sea, being fed by warm waters unimpeded by obstructing continents, was 10–15 degrees Celsius — by modern standards, very high for these latitudes.

Much of Australia was covered with rainforest although there would have been areas which were sparsely vegetated due to their soil type; terrains of granite and, as we have seen, reworked rocks forming steep-sided ridges and infertile plains have existed for most of geological history.

Xeromorphism (plant characters associated with drought resistance) too, had already been around for a long time. Even the very first angiosperms, evolved around 120 million years ago, had small leaves — often considered to be a xeromorphic character — and were apparently associated with the tough environments of seasonal rivers. Possibly these first flowering plants were 'weeds' flourishing in dry, disturbed environments.

That flowering plants may have evolved under regimes of at least some aridity is indicated by the fact that of the 32 families of flowering plants already evolved at the beginning of the Tertiary, one-third now have arid-zone members. Furthermore, on a continental scale extreme aridity has always been associated with cold times or ice ages. Significantly, one of the plant families was the Chenopodiaceae, the saltbushes and bluebushes. These grow in communities which are characteristically low, open and dry.

So, we might imagine inland Australia at the beginning of the Tertiary to be typified by xeromorphic plants growing in open communities on infertile soils, in close proximity to closed-canopied rainforests.

The period of warm seas lasted until 51 million years ago when temperatures dropped sharply, the most acute drop in world temperatures experienced during the entire Tertiary period. It brought ocean temperature into the glacial range for the first time in 200 million years. The reasons for the drop were related to the break-up of Gondwana and the consequent strengthening of the Antarctic Circumpolar Current which up till then had been blocked by land. The current pushed the subtropical weather systems northwards as Australia pulled away from Antarctica. At the same time the Transantarctic Mountains pushed up five kilometres, further catalysing the icy reaction set up by the fragmenting super-continents. Australia was still apparently clothed by a deeply weathered mantle of up to 100 metres deep formed under the conditions of high rainfall. Large rivers meandered over the continent.

This period is believed to be marked by silcrete formation in extensive low-lying areas of the

44 Million years ago

44 million years ago Alice Springs was in Latitude 45° S. Northern Australia had entered the desert latitudes but warm ocean currents allowed adequate rainfall. **Nothofagus** ***forests still dominated the continent but*** **Banksia** ***and*** **Casuarina** ***were more common and the first eucalypts and grasses appeared.***

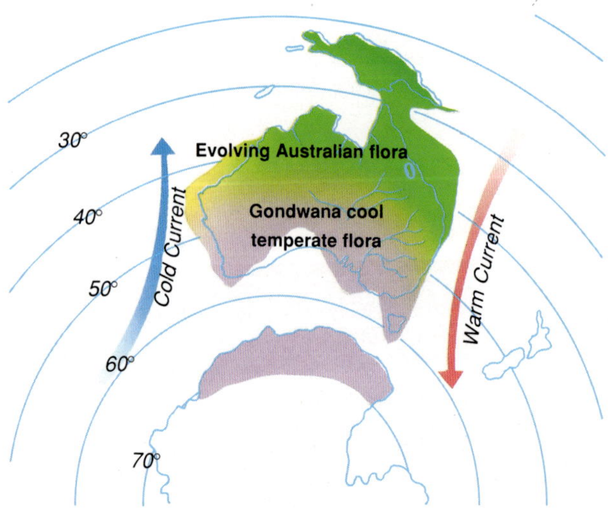

21 Million years ago

21 million years ago Alice Springs had reached Latitude 32° S. The cold circum-Antarctic circulation was well-developed and ice sheets began to develop in Antarctica. While sufficient rainfall still supported woodlands and permanent lakes in the centre, **Nothofagus** ***and rainforests have retreated to the coast and other favourable environments.*** **Acacia** ***and chenopods are found in the fossil record.***

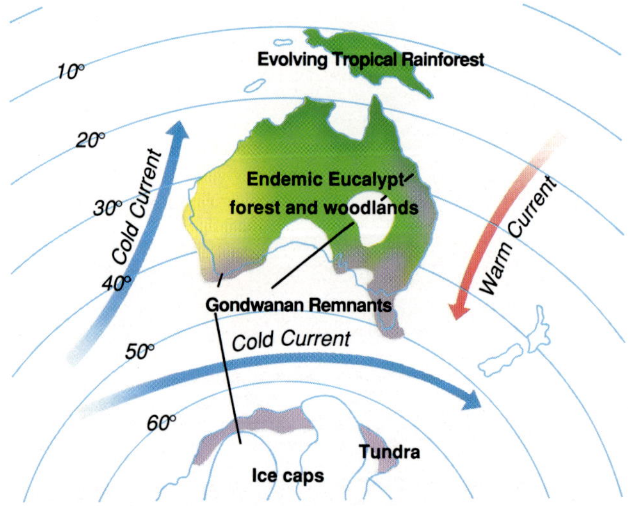

inland: the Canning and Officer Basins to the west and the Great Artesian Basin to the east. With high temperature and rainfall, silica moved through the soil in solution from the basin margins to the lower reaches where it accumulated in the deeply weathered and probably waterlogged material. A cement of quartz or opal was precipitated out of solution. Today this cement forms protective crusts on the flat-topped hills or mesas so characteristic of dry deserts worldwide.

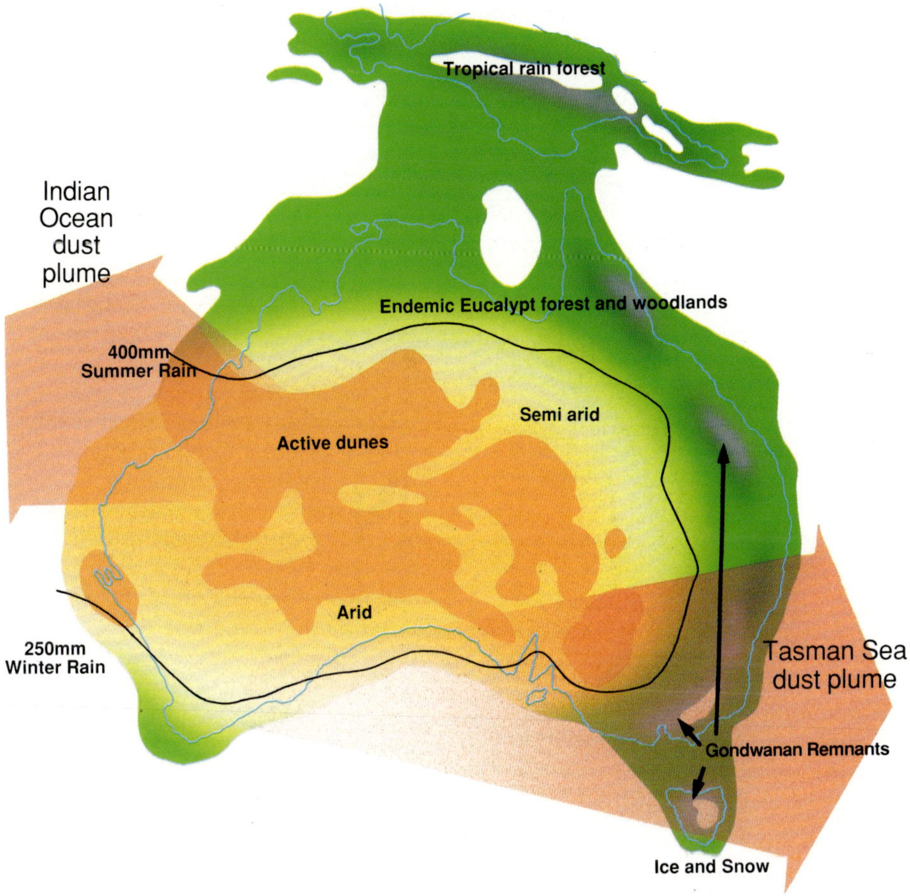

18,000 years ago

The Australia of 18,000 years ago was in the clutches of the last of a series of glacial periods which characterise the present ice-age. Inland Australia was extremely arid, sand dunes were active and dust plumes into the Tasman Sea and Indian Ocean from the central deserts were common. In geological terms this was a period of extinction which continues under European management, despite the improved climate of the last 200 years.

10 Million years ago

10 million years ago the Antarctic ice cap was a single sheet. Circulation patterns in both ocean and atmosphere were similar to today. Central Australia was drying out and was semi-arid. Woodlands, shrubs and grasses dominated the inland.

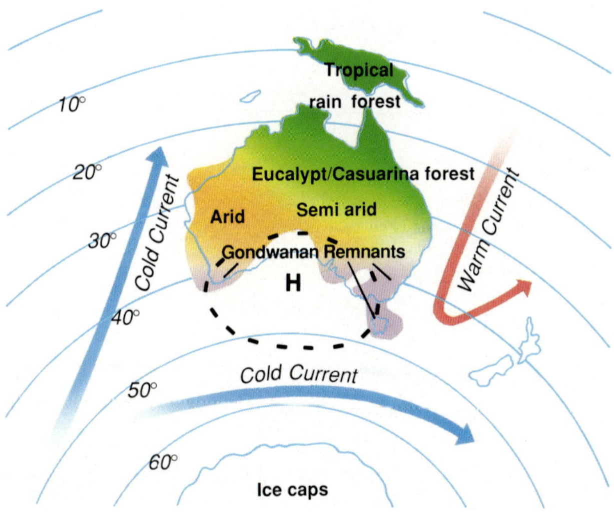

Present day

The Australia of today is in an inter-glacial period. Southern Australia is a region of winter rainfall; northern Australia is a region of summer rainfall. Climatic conditions of the inland have ameliorated since the last glacial period though the rainfall of 250 mm or less is unpredictable. Arid Australia is well vegetated with woodland, grassland and shrubland. Sand dunes are stable.

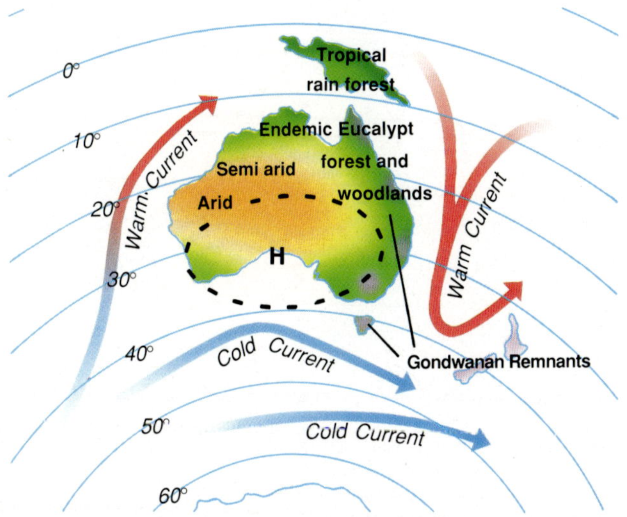

That silcrete formed at this time, however, indicates that the climate was changing from one of constant rainfall to a seasonal regime where at least part of the year was hot and dry. Evidence from pollen also shows that a more complicated mosaic of vegetation was developing in inland Australia in response to both the soils and the seasonally dry climate.

The vegetation graded from rainforest of the type now found in today's tropical regions but at higher altitudes, through to sclerophyllous tree vegetation (mainly *Casuarina*), and some grasslands.

Sclerophylly, or leaf hardness or toughness, although present in all other continents, is recognised as particularly Australian because it is so widespread here. That sclerophylly was already well established is indicated by the abundant *Casuarina* with its thin hard leaves. But rather than having a great deal to do with low rainfall, it is more an adaptation to nutrient infertility; poor soils, like sclerophylly, occur right across Australia.

By the Early Oligocene, about 38 million years ago, ice had formed in Antarctica. And later, at around 30 million years ago, there were periods when glaciers were more extensive than today. It was at this time that the final umbilical attachment between the continents was severed. The cold Southern Circumpolar Current came into its own, sweeping away the moist air of the tropics. In inland Australia there are no fossil records at all for the entire Oligocene; although speculative, this too may reflect relative aridity because fossilisation usually requires damp conditions.

Nevertheless rainforests spread out once again during the early Miocene in response to warm temperatures and high summer rainfall. Strangely, this warming trend was not reflected in Antarctica. Here, spasmodic ice advances, punctuated by warmer periods, had continued since the Late Oligocene. At 14 million years ago, however, there was a sudden drop in temperature and rapid expansion of icesheets in both the Arctic and Antarctic; enough ice expansion, in fact, to mark this as the beginning of the present 'ice age' (rather than the currently accepted 2 million years). The present ice age is characterised by regular glacials interrupted by short interglacials (one of which we are currently in).

The vast rivers that characterised the continent dried up. Temperatures plummeted and with them the capacity of the atmosphere to hold water. Under these circumstances glacials coincided with periods of aridity and the warmer interglacials with higher sea levels and more rainfall.

In Australia the old rainforests were feeling the strain. They were beginning to fragment and become interspersed with *Casuarina* and *Acacia* forests with some *Eucalyptus* and grass and herbs (wildflowers). In deposits near Woomera some macrofossils include fruits with likenesses to several xeromorphic taxa — *Eucalyptus*, *Leptospermum*, *Melaleuca–Callistemon*, *Grevillea* and *Santalum* — and there are some leaves that look like *Banksia*. These were growing side by side with plants which have affinities with modern rainforest plants from New Guinea.

In shady and protected waterholes, trees with rainforest affinities, such as this fig (Ficus *sp.*) *in the foreground, still survive.*

Above, Casuarina *such as this stand of Desert Oak* (Allocasuarina decaisneana) *have been a part of inland Australia since the inland was covered with rainforest. Left, palms hidden in moist pockets of desert ranges such as these* Livistona alfredii *in the Hamersley Ranges hint at a wetter and more luxuriant past.*

Above, Evidence of Homo sapien *— the ultimate ice-age animal.*

Ice-age Animals

Unfortunately the fossil record for wildlife is as murky as the swampy lakes that sprinkled inland Australia. Few records exist until the Miocene, about 20 million years ago, by which time the marsupials — along with the songbirds and land birds and some lizards — had already radiated (evolved and spread) to an extraordinary extent from their Gondwanan ancestors.

At least six different species of koalas and a variety of possums ranging from minuscule to large (cuscus-sized) inhabited the rainforest canopies sheltering a greatly expanded Lake Eyre. Songbirds — ancestors of the robins, scrub wrens, sitellas, pardalotes, woodswallows, butcherbirds, bowerbirds, warblers and honeyeaters — were prevalent. Indeed the structure of the songbird fauna had already begun to resemble its present form.

Several types of diprotodont marsupials, resembling cow-sized wombats, browsed the rainforest shrubs. On the ground smaller marsupials scurried, varieties of rat-kangaroos, bandicoots and insectivorous dasyurids. Ambling and crawling through the litter were types of goannas, dragons and skinks as well as large snakes. In the permanent inland bodies of water were flamingoes, cormorants, ducks, geese, rails, grebes, large crocodiles, turtles, catfish, lungfish and large platypuses. There were even freshwater dolphins in the ancestral Lake Frome area.

It is likely that bats, perhaps Australia's first Asian immigrants, had also arrived.

But by 8 million years ago the characteristic rainforest fauna had disappeared from central Australia, leaving only a few trees able to tolerate or adapt to the arid climate, for example, *Pittosporum*, *Flindersia* and *Ficus*.

The Alcoota fossil site, two hours north of Alice Springs, is a window into this time. The site is a few hectares of pebbly terrain at the foot of a series of silcrete capped mesas. The most common fossil

Only a few species of tree with rainforest affinities, such as this Pittosporum, *were able to adapt to the increasingly arid climate of inland Australia.*

Illustrated by Peter Murray

The largest ever desert-dweller in Australia and, at 3m long and 2m high, the largest marsupial that ever lived was this Diprotodon optatum. *This reconstruction shows a flap-like nose which is a little different from the conventional wombat-like reconstruction. Truly an ice-age animal,* Diprotodon optatum *was the last of the mega-fauna to become extinct perhaps only 10,000 years ago.*

Eight million years ago this tapir-like Palorchestes painei *was a common inhabitant of the dwindling rainforests of central Australia. This diagram shows how palaeontologists cleverly reconstruct the animal from a skull, first tracing in the musculature and then the skin and fur.*

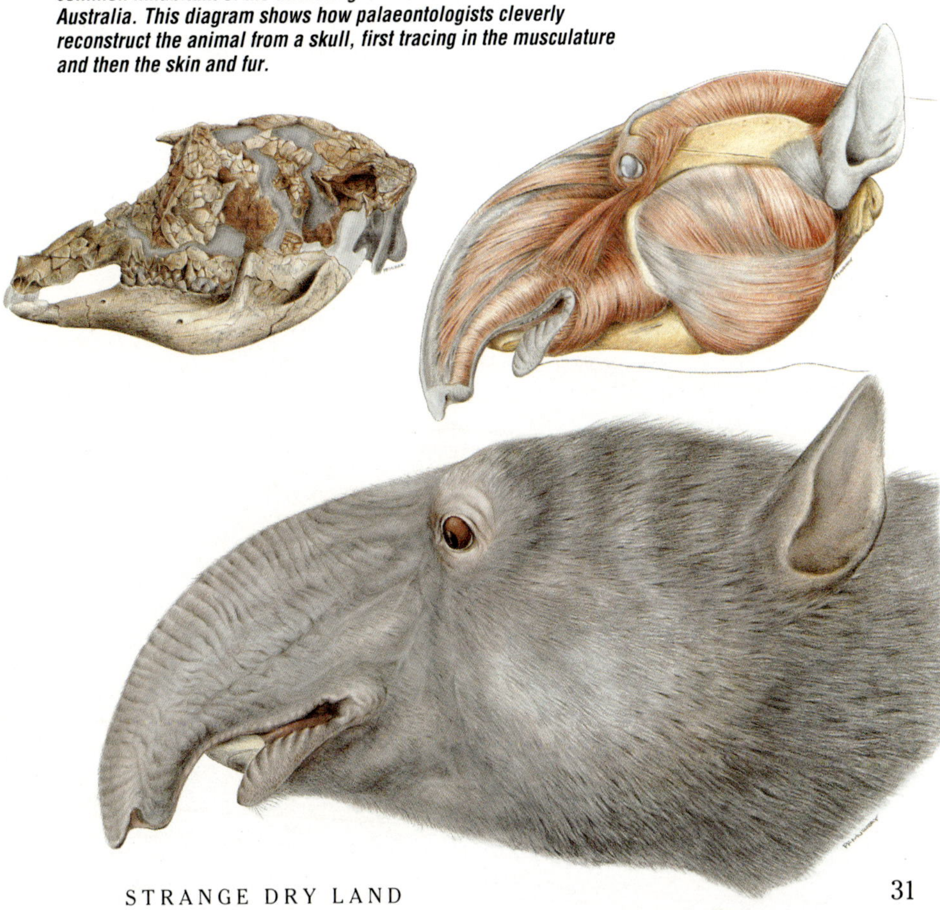

Illustrated by Peter Murray

This two metre-long desert-dwelling python (Morelia bredli) *is a mere shadow of its predecessors which hung in thick loops from rainforest trees in central Australia.*

remains are dromornithids; seven species of behemoth birds weighing in excess of 350 kilograms and standing three metres tall. Because the birds evolved from ground-dwelling ancestors, compared to the marsupials whose ancestors were arboreal, they were more quickly able to exploit the expanding savanna habitat. Nevertheless the diprotodontid marsupials are also abundant at Alcoota, ranging from sheep-sized *Kolopsis* to water-buffalo sized *Pyramios.* In between are hippopotamus-like *Plaisiodon*, and tapir-like *Palorchestes* which apparently had a prehensile trunk.

The predators in this community were primitive thylacines and the marsupial 'leopard', *Wakaleo alcootaensis*. In the lake, aquatic crocodiles thrived. These *Pallimnarchus* crocodiles were large ferocious predators equipped with teeth like serrated knifes. Flamingoes and turtles were also common.

Gone, however, were the many different types of possums, including the peculiar *Ectopodon* possum, which all typified the rainforests of 6 million years earlier around Lake Eyre. In their place came the relative of the first true kangaroos, a sizeable primitive kangaroo called *Hadronomus*.

Rainforests were gradually diminishing at this time and savannas had invaded large portions of central Australia. The large number of terrestrial birds, particularly at Alcoota, certainly indicates open vegetation. The sediments at Alcoota also show clear evidence of alternations between wet and very dry climatic conditions suggesting a monsoonal influence.

The first ancestors of true kangaroos appeared, perhaps in response to the increasing grasslands. And the browsing diprotodontids had increased in size to accommodate a stomach which had to process large quantities of nutritionally poorer herbage as aridity increased.

The glacial–interglacial climatic yo-yo continued through the Pliocene until another precipitous drop in ocean temperature occurred at 2.37 million years ago. In Australia the present climatic zonation of winter rain in the south developed.

The fragmentation of rainforest and expansion of open savanna environments continued, parallelling the abundance, by 3–5 million years ago, of grazing and browsing macropodids. These included the grazing kangaroo *Prionotemnus palankarinnicus* from Lake Palankarinna (close to Lake Eyre) which probably had a similar diet to the modern Agile Wallaby.

The appearance of the bandicoot *Ischnodon australis* at this time represents the first clear adaptation to arid and semiarid grasslands. *Ischnodon* was very similar to the modern-day Bilby (*Macrotis lagotis*) which has a diet in which grass seed is a significant component.

The distribution of the two commonest diprotodontids suggests that the largest, *Diprotodon*, may be a genus that evolved in the arid zone. The dominance of *Diprotodon* in inland Australia in the late Pleistocene seems to have been at the expense of the genus *Zygomaturus*. That arid conditions extended much further than today is indicated by the presence of these species on Kangaroo Island, Spring Creek, coastal Victoria and, of all places, King Island in Bass Strait (it is doubtful that these huge animals could swim expanses of ocean). The time marks the height of the last glaciation and the extreme retreat of the sea exposing a landbridge between Tasmania and the mainland.

DIPROTODON DESERT BEHEMOTH

Some of the mega-Macropodids also show a similar pattern to *Diprotodon* and *Zygomaturus*. *Sthenurus* parallels the range of the inland *Diprotodon*, while the short-faced *Simosthenurus* forms a parallel with the more peripheral *Zygomaturus*.

Diprotodon *dwarfs a Red Kangaroo* (Macropus rufus)*, one of the largest living marsupials.*

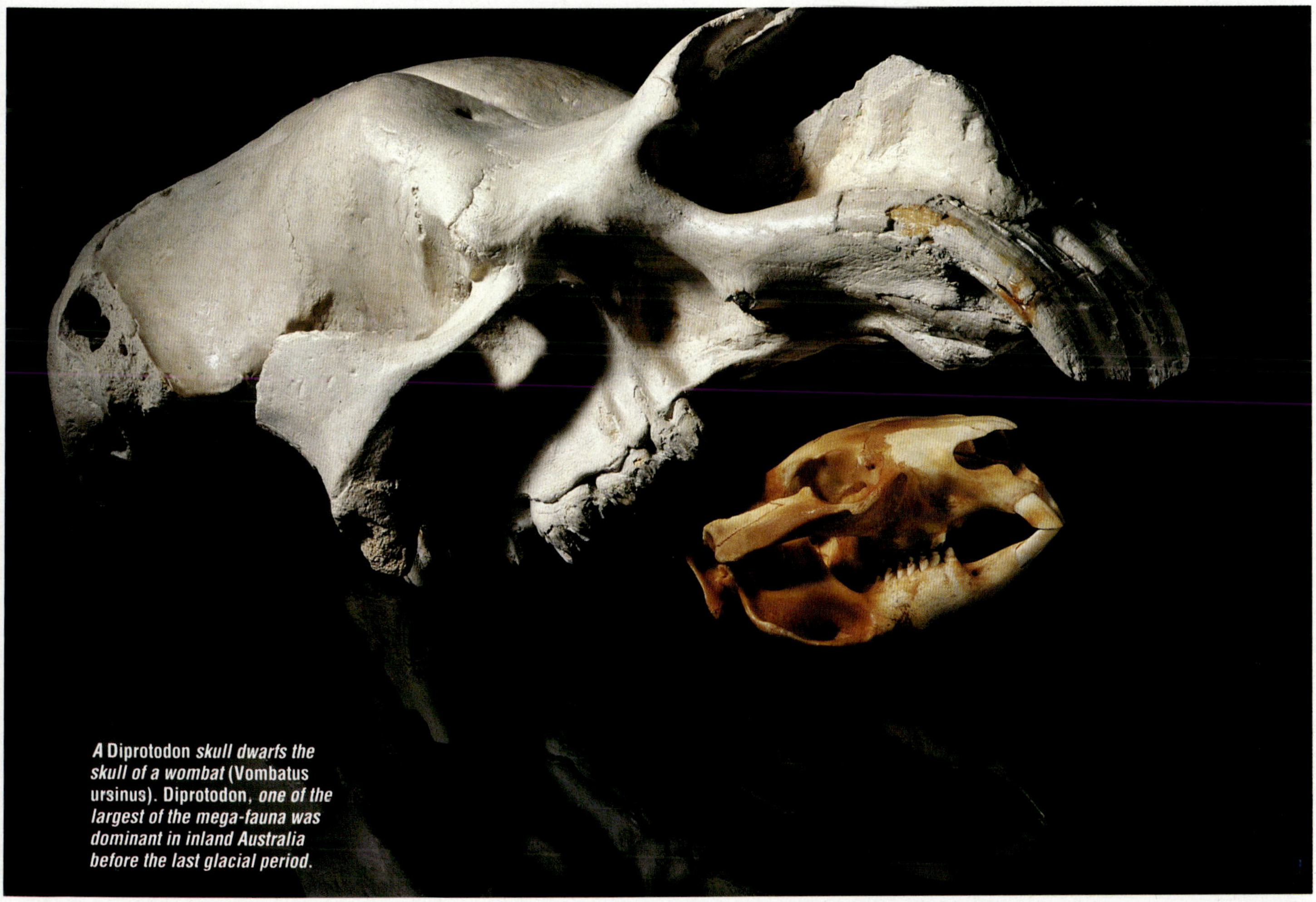

A Diprotodon *skull dwarfs the skull of a wombat* (Vombatus ursinus). Diprotodon, *one of the largest of the mega-fauna was dominant in inland Australia before the last glacial period.*

Living side by side with the Diprotodontids, one of Australia's little-known desert-dwellers, the Bilby (Macrotis lagotis), used to be widespread across the continent until European settlement.

From 2.37 million years ago, but particularly during the past million years, the amplitude of the glacial–interglacial cycles increased. Relatively warm and wet peaks rapidly plummeted to extremely cold and dry troughs, and back again. These extremes were restricted to brief peaks, but the overall cooler temperatures extended over a greater length of time than before the Pleistocene and so produced an overall decline in rainfall in central Australia with a corresponding increase in arid-land vegetation.

Of the land birds, the ancestors of the Emu, Budgerigar, Mallee Fowl, Cockatiel, Little Button-quail, Black-eared Cuckoo and White-backed Swallow were about. These birds are relicts of old evolutionary lines and not related to other Australian birds. The Blue Bonnet, Crested Bellbird, whitefaces, Striped Honeyeater, Pied Honeyeater group, Spiny-cheeked Honeyeater, chats, and four genera of raptors (one kite, one harrier and two falcons) had also evolved by this time within the arid zone.

The continual breaking-up and re-forming of forest patches catalysed the evolution of another 30 genera of birds which, today, have close relatives around the periphery of Australia.

The big-is-better trend continued in the terrestrial animals resulting in the mega-fauna: the large diprotodontids, *Diprotodon* and *Zygomaturus*; the giant wombat, *Phascolonus*; and the colossal kangaroos, *Protemnodon, Sthenurus* and *Procoptodon*. The wildlife at the beginning of the Pleistocene also included small macropodids like today's desert-dwelling wallabies (*Onychogalea* and *Lagorchestes*) and a bettong (*Bettongia lesueur)*. Tasmanian Devils (*Sarcophilus)* and possums (*Trichosurus*) also maintained a strong hold.

The first appearance of rodents in southern Australia is particularly interesting. They appeared in eastern Queensland about 4.5 million years ago, but by 2.5 million years ago they were also found in southern and central Australia. In moving inland they diversified considerably, evolving into forms like the Stick-nest Rat (*Leporillus* spp.), and a Spinifex Hopping-mouse (*Notomys*).

Of the lizard fauna, very little is known. The best fossil lizard remains are those of varanids, the goannas, which have been found from mid-Miocene deposits right up to the Pleistocene mega-goanna, *Megalania prisca*, which grew to five metres in length and probably weighed about 1000 kilograms.

By the late Pleistocene the wildlife — apart from the mega-fauna: diprotodontids, giant kangaroos, wombats and birds — is most notable for its similarity with the fauna of today. Bettongs (*Bettongia*), small wallabies (*Lagorchestes* and *Onychogalea*), bandicoots (*Chaeropus ecaudatus*) and the Bilby (*Macrotis lagotis*) are identical to today's fauna.

Flamingoes, however, died out during the late Pleistocene. Unlike Australia's other endemic water-birds, only flamingoes needed lakes of permanent shallow water. Therefore, it is probably more than a coincidence that they died out at a time which also marked the end of stable shallow lake conditions in central Australia and the beginning of the late Pleistocene regime of extreme climatic oscillations.

This Spinifex Hopping-mouse (Notomys alexis) *is a beautiful little native rodent. It evolved from ancestors which appeared in the fossil record of eastern Queensland about 4.5 million years ago.*

Dune slopes are usually stable, with vegetation covering them. The crests, however, are mobile and wind-sculpted daily. Simpson Desert, west of Birdsville.

Dunefields

During the past million years there have been at least eight glacial–interglacial cycles. In inland Australia glacials are characterised by a drop in average temperature (but summers remain hot), a drop in rainfall (perhaps by as much as 40 per cent), and a dramatic increase in windiness.

Under these conditions dust and sand, stripped from the earth, first become airborne and then pile into dunes. Prior to the present interglacial, the Holocene, dune building has occurred at least eight different times, spanning 370 000 years. Each of these has coincided with glacials.

It comes as a surprise to discover that the largest sand ridge desert in the world — Australia's Simpson Desert — is a young desert formed only during the last glacial maximum, around 18 000 years ago.

The Simpson Desert is the driest region of the driest habitable continent. Summer temperatures reach a withering 50 degrees Celsius and drop to –6 degrees celsius in winter. Average rainfall is 100–150 millimetres and the incidence of severe drought is about 30 per cent to 40 per cent of the time.

The dunefields cover an area of 159 000 square kilometres with individual dunes ranging from 10–35 metres high and up to 200 kilometres, and sometimes more, long. They run parallel to each other for hundreds of kilometres in a NNW–SSE direction with an average spacing of 511 metres. The dunes are asymmetrical in cross-section with a steep eastern slope of about 20 degrees and a gentler western slope of 12 degrees. While the dune slopes are stable with vegetation covering them, the crests are mobile and are wind-sculpted daily. The dunes are part of a continental anticlockwise swirl of longitudinal dunes broadly approximating the winds of the subtropical anticyclone systems which sweep across the continent. Today only the very southern parts of the desert get strong enough winds to shift much sand.

Above, a shrub disrupts the pattern created by wind-rippled sand near Uluru.

Bare dunes such as this one looming over an Acacia ligulata are not the norm in Australia's dunefields. They occur in areas such as the very southern Simpson Desert, where winds are strong enough to shift the sand.

With its great sandy ribs, the Simpson fits the classic desert image. It is, however, a relatively shallow sand desert (as far as sand deserts go). And, although it grew during the last glacial period, its birth was initiated during a much wetter time than now when watertables were at or above the ground. Today the watertable in the region is about 40 metres below the surface.

About 50 000 years ago, during the last interglacial, Lake Eyre and surrounding lakes were full. The large Channel Country creeks such as the Cooper looped and twisted towards these central lakes, carrying mainly fine-grained pale sand from Cretaceous and Jurassic sandstones and mudstones. Further to the north the Hale, Hay and Plenty rivers were depositing broad tracts of red sand torn from ancient low ranges, such as the MacDonnells, to the north and north-west.

EARLIEST ARIDITY IN CONTINENTAL CORE

The 500 kilometre chain of salt lakes in central Australia known as the Lake Amadeus Salt Lake System is the epilogue of a story going back perhaps 12 million years.

One hundred metres of sediments underlying the system, like growth rings, show the history of the surrounding landscape. The bedrock hills Uluru (Ayers Rock) and Kata Tjuta (the Olgas) protrude through these sediments.

The basal unit is a clay known as Uluru clay and was deposited in shallow lake and river environments. The environment was wetter than it is today although the sediments show that there were frequent, perhaps seasonal, dry periods and that the lakes were periodically saline. During this time of clay deposition there were no sand dunes and a more permanent and more extensive vegetation existed than exists today.

Around one million years ago the first evidence of dry conditions like today appeared. The evidence of dunefields at this time is, in fact, the earliest Australian dunefields recorded.

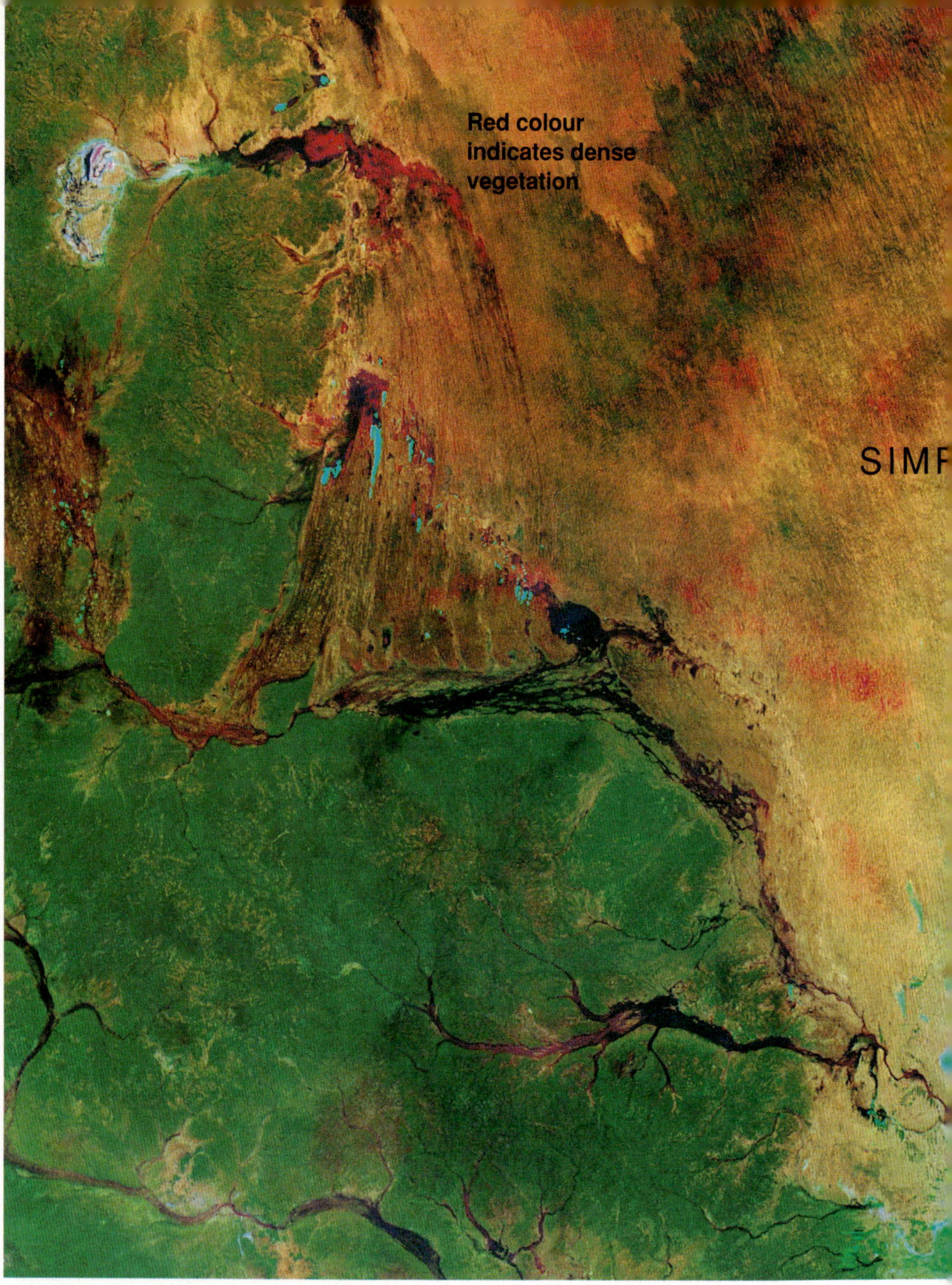

From about 30 000 years ago the lake levels began to fluctuate in the southern Simpson Desert region as temperatures worldwide began to plummet before the impending glacial. Cooper Creek changed from a series of vigorous anastomosing streams to a less sinuous system beginning to deposit mud on floodplains and in the evaporating lakes. In the increasingly windy conditions crescent-shaped dunes (or lunettes) containing mud pellets were heaped around the downwind shores of lakes, like sand dunes on a beach. Longitudinal dunes streamed like hair further downwind from these fringing dunes, blanketing and smothering creeks in the gale-force winds of the time.

Stranded saltpans, thus formed, contributed to the dunefield by the same process. The entire landscape of groundwater evaporation pans, lunettes and the characteristically pale-coloured dunes is called a 'boinka'; a landscape formed by saline groundwater discharge and evaporation. In particular, boinkas were formed about 23 000 years ago when the watertable oscillated just above and just below the ground surface under intense seasonal conditions of hot relatively wet windy summers and cold windy winters. Each cycle resulted in a deposit of mud. When the lakes or pans were seasonally dry the mud would be swept, or 'deflated', from the saltpan surface into lunettes and then, from these, into longitudinal dunes.

The dunefields to the north were formed in a different way and were not groundwater controlled. During the same high wind conditions, alluvium deposited by the northern rivers was stripped from slightly elevated areas (contrary to the way in which boinkas were formed from the sediment of depressions) and then heaped into dunes. These dunes are brown-red because the sediment from which they are made came from a different source than the pale dunes of the southern Simpson and were not affected by saline groundwaters or 'diluted' by mud pellets.

The Simpson Desert lies in a huge basin into which waterborne sediment has spilled for millions of years providing the sand for this globally significant dunefield. The Great Sandy dunefield and its

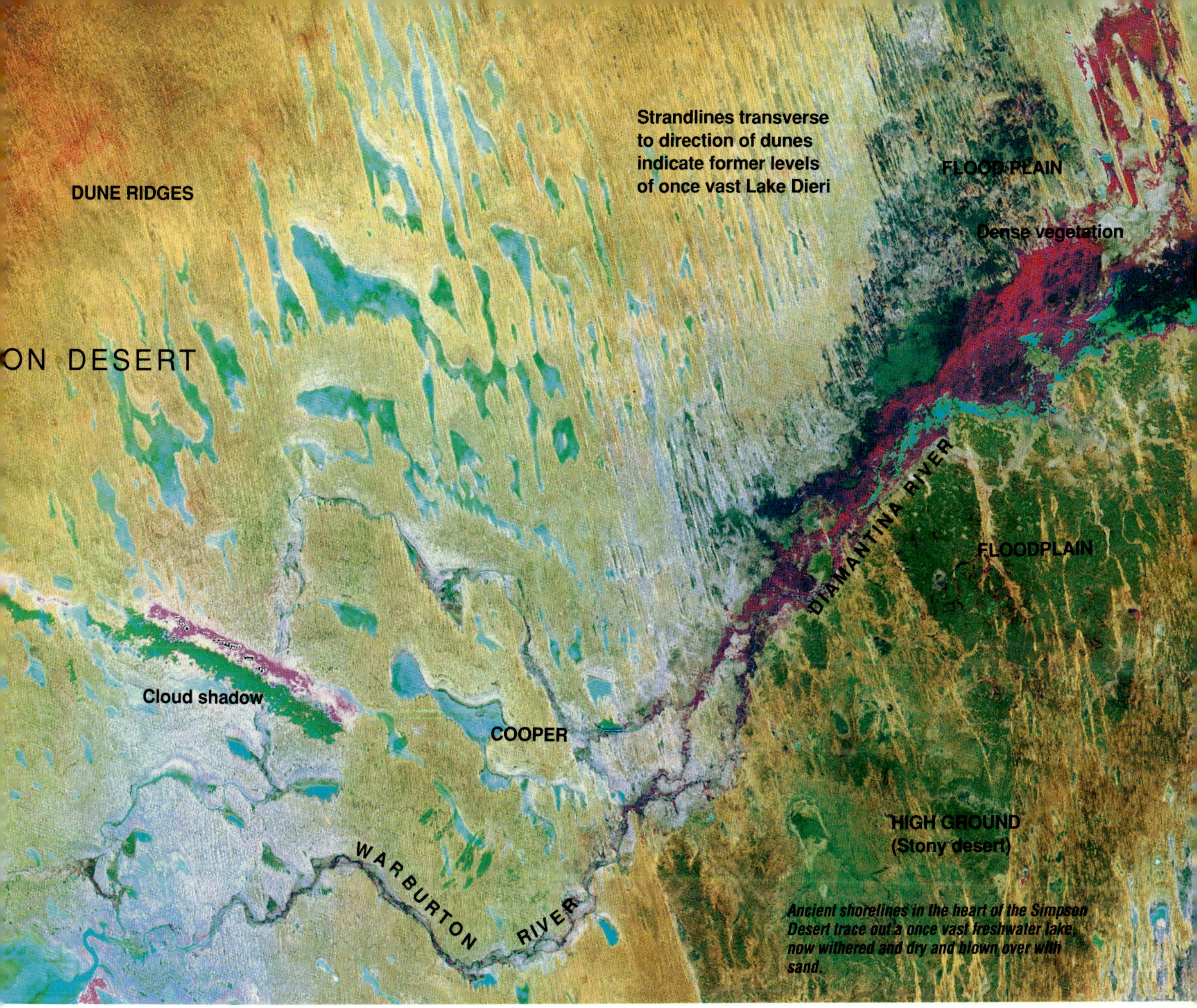

Ancient shorelines in the heart of the Simpson Desert trace out a once vast freshwater lake, now withered and dry and blown over with sand.

extension into the northern Gibson dunefield also lie in topographic basins, but much shallower ones than the Simpson. Their main source of sand has been the palaeorivers around which linear dunes also stream downwind from areas such as Lake Auld in the Great Sandy dunefield. The Tanami dunefields and those of the southern Gibson Desert and Great Victoria desert all lie on low ground in topographically more complex areas scattered amongst low ranges, dissected tablelands and plains.

During the peak glacial times, groups of plants and animals which lived in the arid zone were assailed by the wind and the sand-swept desiccated environments that typified the time. Many became extinct but others would have been held under siege in mountain fortresses, like islands in a sea of hostility, or banished entirely to the periphery of the arid zone. Other plants and animals evolved to exploit the expanding sandy environment.

Even at the peak of glacial periods there was probably never a barren wasteland, although rainfall would certainly have been less and the environment much more difficult than that of today.

During the last glacial around 18 000 years ago, the Australian landmass was about 25–30 per cent larger than it is today. This led to fewer rainfall systems, particularly the summer monsoons, penetrating into the interior. Simulation of this climate by scientists has put estimates of the average annual rainfall of the interior at 50 per cent of modern averages. The plains surrounding Lake Eyre today — the driest part of the arid zone — come close to what glacial environments might have looked like: a patchy scatter of perennial shrubs getting by, creases of dark woodlands in creeklines and ephemeral wildflowers and grasses after rain.

Winter temperatures would have plummeted, dragging down the average yearly air temperatures across the interior by perhaps 5–6 degrees. This is despite a small increase in summer temperatures. Plant species affected by increased frosts or requiring summer rainfall would clearly not thrive, although major vegetation formations already adapted to wide fluctuation of rainfall would have persisted.

The New Age

The breaking-up and re-forming of habitats during the late Pleistocene climatic waxing and waning was a potent mechanism for the generation of new species.

Lake Mungo, one of the oldest human campsites known, covered with sand. Aboriginal occupation of Australia predates the formation of our biggest dunefields, such as the Simpson Desert.

This Chestnut-crowned Babbler is the most endearing and querulous of birds. Its evolution coincides with the recent development of dunefields and sandplains.

Lizards in particular radiated explosively and became associated with specific habitats, particularly with the expanding sandridges and sandplains and with *Acacia* shrublands. Bottlenecks between these habitats constantly changed, alternately isolating populations of sandridge and sandplain species during moister periods and then *Acacia* shrub lizards during the dry periods.

Today, for instance, a population of the skink *Ctenotus brooksi* in the Simpson Desert is cut off from its relatives in the Great Victoria Desert and the central ranges by a narrow stretch of scrub in the south-central Northern Territory. This barrier is probably no older than a few thousand years but already a distinct subspecies of *Ctenotus brooksi* has evolved.

The Knob-tailed Gecko *Nephrurus laevissimus* also has an isolated population in the Simpson Desert.

Of the shrub-preferring lizards at least eight species of dragons, geckos and skinks have crossed from the deserts of the west to the deserts of the east through the 'Giles Corridor'. This corridor of *Acacia* shrubland connects the Pilbara region to the central ranges by way of the Lake Carnegie region in the Great Victoria Desert and the southern part of the Gibson Desert. The Giles Corridor is the only continuous ribbon of shrublands through the Great Victoria Desert, which is dominated by sandridges and sandplains.

Even birds that would normally fly over these lizard-barriers have been affected by the development of dunefields and sandplains in arid Australia. One of the best examples is in three species of the ground-dwelling quail-thrushes (*Cinclosoma*). The Chestnut Quail-thrush is an inhabitant of sandplains and dunes. But this same system of dunes which connects the western central and eastern populations of Chestnut Quail-thrush severs the two main populations of the Cinnamon Quail-thrush: one population lives in the low shrublands of the Lake Eyre Basin and the other in a similar habitat on the Nullarbor. They are effectively separated by a band of dunes that rims the Nullarbor Plain from the Great Victoria Desert through Barton to the top end of Eyre Peninsula.

The Chestnut-breasted Quail-thrush evolved from the Cinnamon Quail-thrush. The events which led to the divergence are not clear but they are thought to have involved the 'Eyrean Barrier'. Since the Pliocene this shallow trench, apparently caused by downfaulting and intermittently filled with water, is said to have cut inland from the head of Spencer Gulf to the Lake Eyre–Frome Basin separating desert fauna into western and eastern parts.

Recent geological evidence, however, has shown that no such structure ever existed. Science is an exciting and challenging area with many ideas and hypotheses being launched and then, more often than not, shot down. Perhaps the development of dunefields and sandplains alone are enough to explain the speciation of the quail-thrush? If not then the speciation of other birds such as the babblers *Pomatostomus* spp., the most endearing and querrulous birds, now provides a dilemma. Today the White-browed Babbler overlaps the Chestnut-crowned and meets the Halls Babbler without interbreeding. Originally, however, the three babblers are all thought to have formed by a series of separations and overlaps controlled by the Eyrean Barrier in this way: An early separation of the White-browed Babbler from Western Australia formed the Chestnut-browed Babbler; the drying of the Barrier allowed both

FACTORS DETERMINING DESERT DUNE TYPE — THE SHALLOW SIMPSON DESERT

In the world there are four elemental types of desert dunes: longitudinal, transverse, barchan and star. The wind regime, the volume of sand available and the amount of protective vegetation all determine which dune type will form. Results of extensive analysis of these factors have shown that transverse dunes occur where sand is abundant and the wind is moderately variable; star dunes occur where sand is abundant, vegetation is absent and wind variability is at a maximum; barchans occur where there is very little sand, little vegetation and almost uni-directional winds; and longitudinal dunes occur where the winds are a little more variable but sand supply is limited. So the sea of sand that is the Simpson Desert, the largest sandridge desert in the world, is actually a shallow desert with only a thin veneer of sand.

Linear or longitudinal dunes, such as those found in the Simpson Desert, with vegetation on their flanks.

A transverse dune usually consists of plenty of sand and the wind is moderately variable.

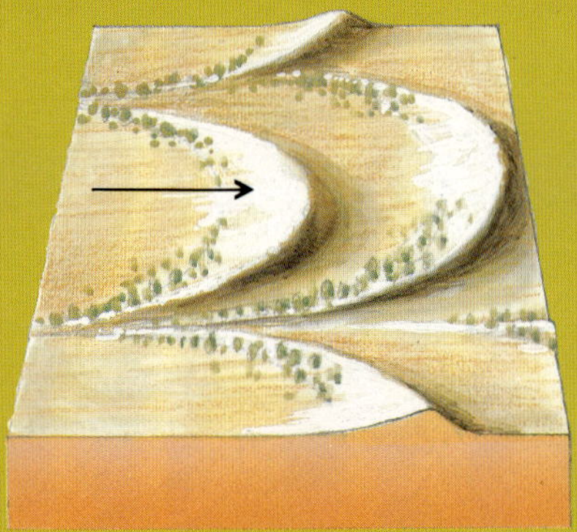

Parabolic and barchan dunes are the sort you often find in blowouts on the coast. The trailing 'arms' are stabilised and slowed by vegetation.

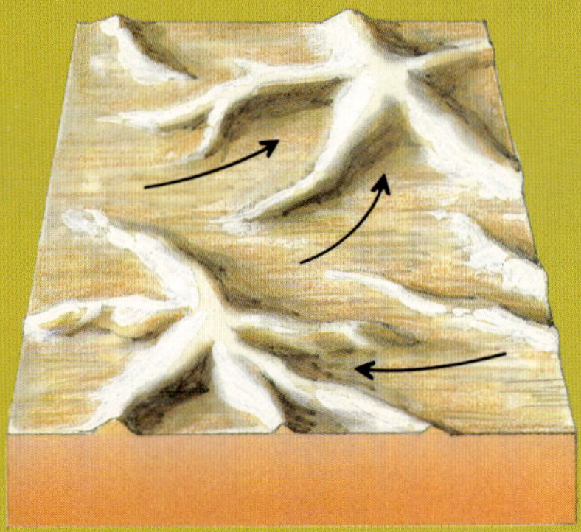

Star dunes form where sand is abundant but they support no vegetation.

populations to overlap; this overlapping population formed the Halls Babbler after a further separation. Without the Eyrean Barrier other mechanisms will have to be sought to explain this speciation.

Perhaps because the wind consistently blows from the west, there is a common tendency for western birds to overlap eastern rather than the reverse (it is much more difficult to flap against the wind). This is particularly evident in populations of Ringneck Parrot and Grey Shrike-thrushes in the Flinders Ranges, where only western parental forms exist.

However, these desert ranges themselves are perhaps the primary source of divergence and radiation within the desert avifauna. The Hamersley and central ranges — including the MacDonnells — have been particularly important, especially during the last phases of aridity. Subspecies of Spinifex Pigeon, Ringneck, Striated and Dusky Grass-wrens and Western Bower-birds for instance probably diverged then. When the climate ameliorates these birds spread out to overlap with their relatives. This explains why the Red-capped and Scarlet Robins, the Variegated and Blue-breasted Fairy-wrens and the Grey-fronted and Yellow-plumed Honeyeaters occur together today.

The desert mountain ranges also acted as refugia for some plants. This is certainly true for *Acacia*. The richest concentration of wattles in the arid zone, in fact, is found in the rugged uplands of the Pilbara and in central Australia. Forty-six species occur in the Hamersleys and in central Australia 28 of the 32 endemic species predominate in the rocky ranges and tablelands. A major clustering of saltbush or *Atriplex* species also occurs in arid hills, as does *Cassia*.

The arid mountain ranges not only act as refuges but also as reservoirs of currently evolving species. Perhaps the best example of this is in the jumble of hills, flats, breakaways and gibber plains on the ancient shield region of Western Australia. Here *Eremophila* is undergoing rapid evolution into a series of complexes and inter-grades. The Hamersley Ranges are also a region accumulating recently

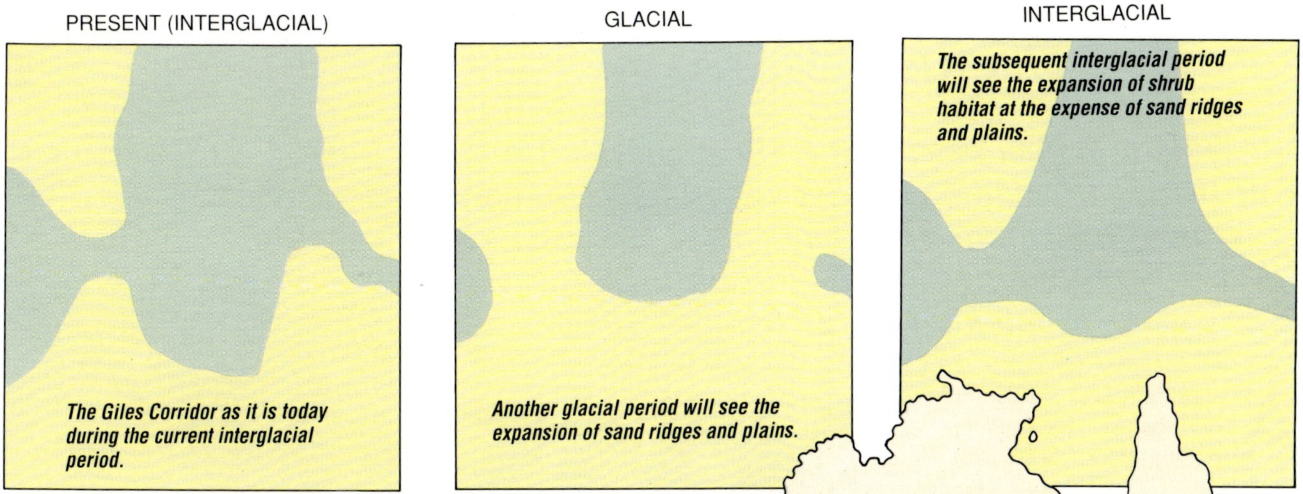

The Giles Corridor as it is today during the current interglacial period.

Another glacial period will see the expansion of sand ridges and plains.

The subsequent interglacial period will see the expansion of shrub habitat at the expense of sand ridges and plains.

Shrublands

Spinifex/sand

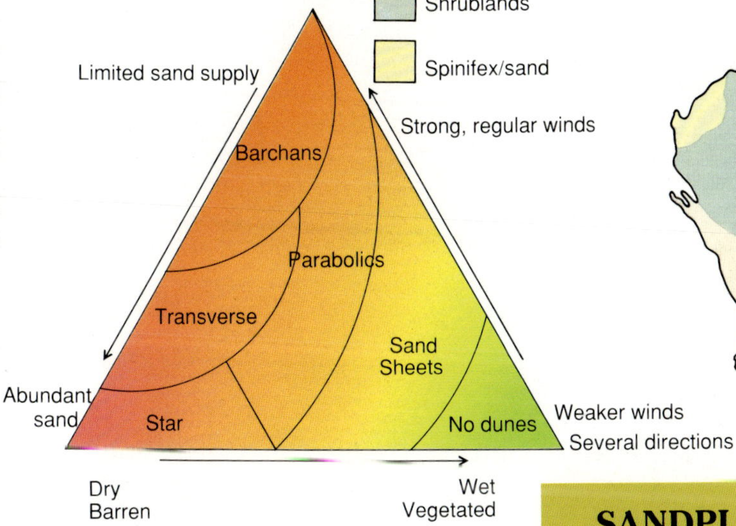

Factors controlling sand dune shapes: abundance of sand, wind strength and direction of vegetation.

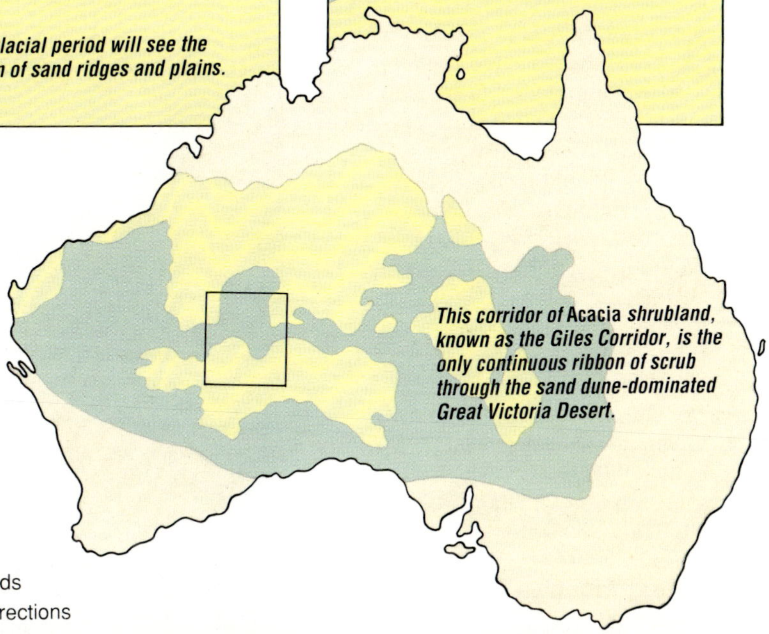

This corridor of Acacia shrubland, known as the Giles Corridor, is the only continuous ribbon of scrub through the sand dune-dominated Great Victoria Desert.

derived species of *Acacia*. Of the 46 species occurring here, 44 have close relatives from which they are recently evolved.

These reservoirs of species occasionally flow out from the ranges to populate the environments of the plains. For example, two species of *Eremophila* have left the scatter of hills on the Western Australian shield to occupy the surrounding sand dunes. In central Australia, while 28 of the 32 species endemic to the region remain in the rocky region, four of the most labile species have spilled out onto the expanding dunes and plains. Even the species of spinifex which are widespread in the sandplains and dunefields probably evolved in the rocky regions where they are also found.

SANDPLAIN FLORA

Research into the distribution of desert plants has revealed a remarkable number of species with disjunct distributions including several species of grass, mistletoe, *Acacia*, *Eucalyptus*, *Eremophila*, *Solanum* and *Cassia*. Most of these species probably had a wider, more continuous distribution which has been broken up as recently as 20 000 years ago by the sandy areas which these species have not been able to recolonise. The sandplain flora is therefore a youthful flora composed of (i) species invading from the surrounding stony plains and desert mountains, (ii) species migrating from other sandy areas such as the coast, and (iii) species recently evolved within the sandy areas. Evolution has also continued within the separate populations and in some cases populations have already diverged enough to have formed new species.

Ice-age Man

From a human perspective, the productivity of the land would be considerably lower. Many central Australian trees and shrubs are frost sensitive, including some important bush tucker plants such as figs.

A general reduction in plant cover would have reduced the abundance of medium to small mammals and regular access to much of the region was effectively denied because of the lack of standing water such as small wells, soakages and small ephemeral rockholes. Only springs from regional aquifers within the desert uplands might have been permanent.

Yet, people did survive in the very centre of the continent throughout the last glacial maximum. This is shown by the archaeological evidence from Puritjarra Rockshelter in the Cleland Hills of central Australia which shows a more or less continuous occupation of the centre from 22 000 years ago right through the glacial period to 13 000 years ago.

Human ecology during the last glacial maximum must have been characterised by very opportunistic use of a wide range of resources including reptiles, small flock birds, small mammals and roots and bulbs. A complete absence of grindstones from Puritjarra at this time confirms that intensive use of seeds as a staple, which characterises later periods, was not a feature of Pleistocene subsistence in the arid interior.

In this less productive time the human's foraging territories needed to be larger, but at the same time people were environmentally 'tethered' to waterholes. Only in the major desert uplands would the distribution and spacing of waterholes allow access to large blocks of country. In the central ranges a population of about 500 people (about one-tenth the density at the time of European occupation) may have been able to survive the last Pleistocene glacial.

Fruits such as these frost-sensitive figs may not have been available to people during the last glacial period.

Recent Aboriginal stone tools: a complete absence of grindstones 20 000 years ago, such as the one in the centre of this picture, confirms that intensive use of seeds was not a feature of Pleistocene subsistence. Human ecology was probably characterised by a very opportunistic use of a wide range of resources.

Under the harsh conditions of 20 000 years ago, these temporary reservoirs may have been useless. People were probably water-tethered to permanent pools in the desert ranges.

Ochre pits, west MacDonnell Ranges.

THE DINGO

'Yellow dog dingo
How old are you
Where do you come from
What do you do'

from a song by
L. King, Australian folk singer

Long ago, perhaps even before 4000 years, an Asian navigator in his hand-hewn boat came to the northern coast of that great southern land inhabited by a fiercely independent dark race of people. The wayfarer had with him his tame wolf or dog. For some reason, perhaps as trade or perhaps because of a shipwreck, the dog remained to breed with other dogs brought in the same way across the sea from Asia to Australia.

It has long been assumed that Aborigines brought the dingo to Australia around 5000 years ago. But evidence recently come to light points to a scenario like the one described above as being more accurate.

Dogs first appear, presumably as domesticated animals, 5500 years ago in north-east Thailand; at 5000 years in northern Vietnam; in Sarawak after 4500 years; in Timor between 3500 and 2500 years; around the same time in Australia (the oldest surely dated fossil dingo remains are 3500 years before present); and in the Pacific islands about 2750 years ago. These dogs, evolved from the Indian Wolf perhaps 7000 years ago, are very similar to the Australian dingo. In fact uncontaminated (by domestic dogs) Thai dingoes, today, are more closely related to Australia's dingo than to domestic dogs.

The spread of the dingo marks almost exactly the spread of people speaking Austronesian languages throughout island south-east Asia and into the Pacific. Dogs may have been taken as live food cargo since dog meat is commonly eaten in parts of Asia today. There may have been cultural reasons too. In Borneo, for instance, a dog is buried with its owner at death so that together they can commence the journey into the afterlife. Perhaps so too did dingoes accompany Asian voyagers to sea, spreading north into China and as far west as Israel.

Different cultures selectively bred their dingoes for different attributes such as short legs or long legs, floppy ears or pricked ones. In this way the domestic dog evolved. The Aborigines did not have this penchant and, as a result, the Australian dingo has remained virtually pure.

Land of Unpredictability

The Finke River, used to trace the geological history of central Australia, also represents the all-important flow and redistribution of water across the desert landscape.

Gorges near Kings Canyon in central Australia harbour tens of rare and relict plants. As much as a third of the entire flora of central Australia can be found in rich areas such as this.

Deep within shady gorges, reliable water and nutrients create oasis-like environments. Bungle Bungles, Western Australia.

In 1894, Professor Ralph Tate, while on the Horn Scientific Expedition with Baldwin Spencer, searched the headwaters of the Finke River for '. . . a vast mountain system capable of preserving some remnants of that pristine flora which had existed on this continent in Paleocene times'. He found that, to the contrary, the inclined slopes and summits of the MacDonnell Ranges were devoid of soil and had 'the appearance of a road newly made with large metal, amongst which porcupine-grass grows usually so densely that progress is extremely difficult and even painful if the pedestrian's legs are not well protected'.

He did find a varied flora; not the 'beech . . . oak, elm or sycamore' that he was looking for but a more peculiar flora composed of 'novelties'. It was found where water concentrated and flowed, deep within the shady gorges and spilling on to the flood-outs of the ranges.

Tate wrote down, for the first time, the concept of areas of 'refugia' — islands of favourable habitats created by a combination of topography, water and nutrients to account for these novelties. He had hit upon an idea that more than any other underpins the way ecosystems function in arid Australia.

The most reliable thing about rain in arid Australia is its unreliability. Yet reliable water is the single most important factor triggering plant growth in deserts. If rain is the orchestral composer of the Australian deserts, then topography is the symphonic conductor redistributing the water that flows across the deceptive uniformity of the inland; in even the flattest environments, water run-off and run-on occurs creating the rich patterns of vegetation and soils, like Aboriginal sand paintings, that can be best appreciated when seen from an aircraft.

The low ranges in particular harvest what water falls and allow for the natural irrigation of plants in these favourable areas at times when the average water availability across the whole landscape would not support growth.

Where water flows, nutrients follow so that plants with better nutrient status will grow in these naturally irrigated areas. If there are no disturbances like grazing or fire, perennials will gradually dominate. These plants and the animals which live on them will have specific life-history strategies adapted to a relatively favourable and reliable habitat.

The Finke River spilling out of the central ranges is itself a ribbon of refuge habitat. The permanent watertable, small gullies and waterholes along its length provide a reliable habitat for birds, fish and plants. Water supplies can be more reliable either by being continuous, as with permanent watertables, or by being relatively regular, as with run-on areas such as the flanks of the ranges.

Further afield, in the broad sweep of the landscape, water supplies are intermittent and perennials with a greater tolerance for drought, such as mulga (*Acacia aneura*), will survive on deeper well-drained soils, and various types of spinifex on skeletal and sandy soils.

Like an old bushman who has lived all his life on damper and salted beef, with few nutrients, mulga and spinifex are tough and wiry and any growth is hard won. If herbivores literally cannot stomach the poor-quality vegetation, for detritivores, particularly termites, it is a banquet. If termites are the grazing animals of the mulga woodlands and spinifex grasslands, then lizards must be the hunters, and it is in these 'poor' areas that we have the greatest diversity of lizards in the world.

Out of the desert ranges and onto the plains, the Finke River sweeps through a country characterised by flat-topped hills and desert sands.

But while water and nutrients go together, what happens when there are nutrients but no water? As the Finke snakes its way across the desert plains to Lake Eyre, its big bends eventually sweep into a country characterised by flat-topped hills or mesas which are crumbling remnants, 'break-aways', of the old deep-weathered Tertiary surface; a flat country becoming flatter. Here, the protective silcrete capping of mesas, now the highest part of the landscape, were originally formed in the lowest part of the landscape, underneath lakes or valleys. The stoney or gibber deserts are the scattered scraps of silcrete left after the planing of even these inverted plateaus. The soils of this landscape are nutritionally rich, being formed initially when the Cretaceous sea swamped most of the continent and subsequently from rich lake and river sediment.

But today these areas are amongst the driest in the continent and are characterised by saltbush and bluebush. While sparse, however, these plants are nutritionally rich. Rich enough to support small populations of native herbivores so that even the sun- and sand-blasted gibber areas have their own characteristic suite of animals.

It would be hard to imagine an evocative landscape like this cut by Amazon-sized rivers. Yet near a large loop of the Finke aptly called Horseshoe Bend, a saltbush flat is

bisected by leafy coolabah trees (*Eucalyptus microtheca*). These are the telltale signs that an ancient drainage line lies buried beneath. Karinga Creek is one of many palaeodrainage channels which once formed vast networks across much of the arid zone. Today some of the interconnected strings of salt lakes and playas across the arid zone mark the courses of these fossil rivers. In the Great Victoria and Great Sandy deserts the palaeorivers are no older than the Cretaceous Sea: about 100 million years. Some palaeorivers on the old shield areas were not inundated by this sea and may date to the Permian ice age 250 million years ago.

Since the mid-Miocene the rivers stopped flowing regularly and over the past 14 million years have been fragmented and drowned in their own sand. Palaeorivers have been so disintegrated that they were only recognised as ancient river courses from space. From here Landsat imagery resolved a vast network of limestone rimmed depressions, strings of playa lakes and claypans into clearly connected palaeodrainage systems. Today these palaeodrainage systems are still active with water slowly percolating through the soil to the lowest parts of the system, the salt lakes. These salt lakes, including the vast Lake Eyre Basin and the Queensland Channel Country are, perhaps deceptively, also refuges; rich areas in the desert landscape where water and nutrients are pooled.

The Lake Amadeus Salt Lake System was once a major tributary of the Finke but today it is a place of blue mirages with only a tenuous connection with a river of sand which, from Horseshoe Bend, coils south-east like a giant oscillating steel spring disappearing into the Simpson Desert and the heat haze.

SLACKWATER DEPOSITS

Slackwater deposits in rivers are formed when sediment-loaded floodwaters are diverted from the main flow to quieter areas where the load settles. Like growth rings, they leave a remarkable record of prehistoric and historic floods. The flood record of the Finke River, as it passes through Finke Gorge, is the best in the country. This is due to an ancient 30 million year old gorge, no longer in use, which intertwines with the main gorge and provides sheltered sites for slackwater deposits to settle (see diagram on p.25, bottom left). This fossil gorge, which is obvious from the air, is thought to have formed during a much wetter time than now. The current more entrenched gorge is perhaps only five million years old.

Carbon dating of these Finke River slackwater deposits has shown that the largest floods for 700 years were in 1967, 1972, and 1974. The fact that these are clustered in the twentieth century indicates that the Finke may be telling us something about world climates. The most ominous explanation is that changes in rainfall patterns reported in Australia and ascribed to the build-up of atmospheric carbon dioxide not only involves shifts in the average rainfall but also involves increases in the size and frequency of extreme rainfall events.

LAKE DIERI

Inland Australia is a vast, flat place. And features are often so large that they are only recognisable from space. The advent of satellite imagery has allowed geomorphologists to recognise immense features, such as the fossil rivers or palaeo-drainage systems once sweeping the continent, and the possible Pleistocene mega-lake which once subsumed all of Lakes Frome, Callabonna, Blanche, Gregory and Eyre. Lake Dieri (Dieri is the tribal name of the Aborigines who once occupied the Lake Eyre region) was thought to have covered as much as 110 000 square kilometres, more than six times the area of the present lakes combined. From space (see p.38) the myriad of large saltpans form a series of arcs, like tide-lines, roughly parallel to the eastern and northern shorelines of Lakes Frome, Callabonna, Blanche, Gregory and Eyre. They appear to be remnants of a gradually diminishing but once vast lake, a veritable inland sea, once occupying the centre of the continent.

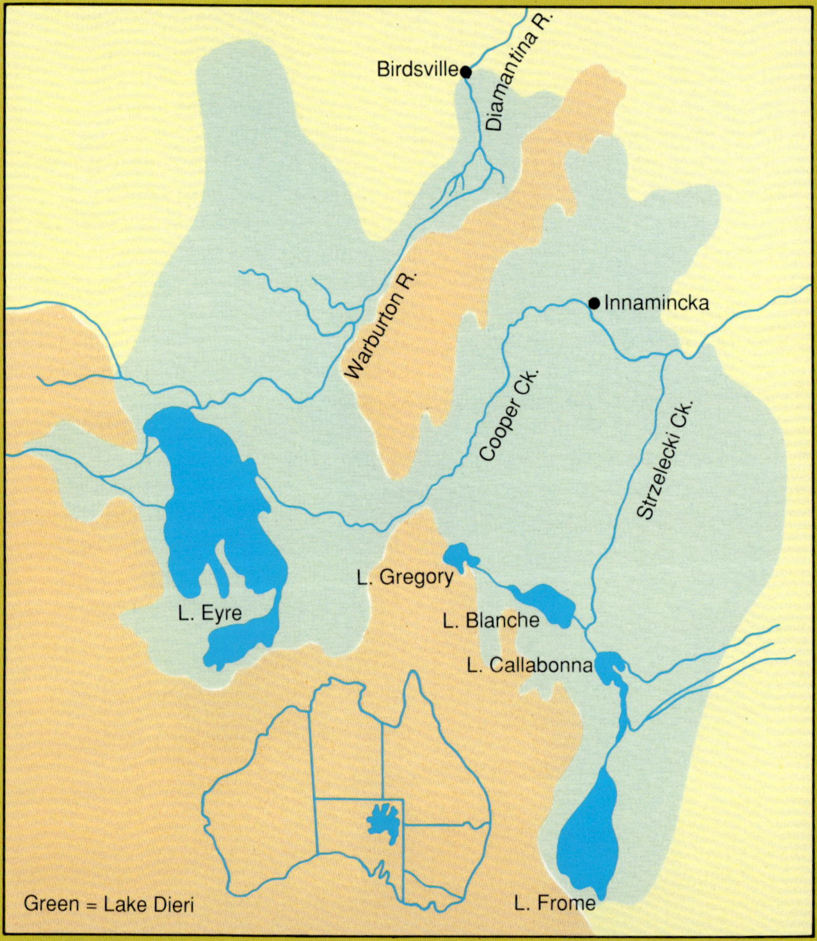

The strings of salt lakes lacing inland Australia are the remnants of forgotten rivers and lakes, full as recently as 50,000 years ago.

Pristine coastal dunes of the Great Australian Bight. Formed during modern times, these are some of the most arid areas in Australia. On the northeast margin of the Nullarbor, inland of these dunes, are perhaps the oldest dunes in the world. Dated at around 35 million years, they trace an ancient shoreline. Their preservation implies a degree of aridity, at least for this part of the continent, since then.

WHY NO CACTI?

In most of the world's deserts succulence, such as occurs in the 2000 or so species of cactus, is a characteristic plant feature. Not so in Australia. Here we have only about seven true stem succulent plants. Succulents, while desert plants, nevertheless require a reliable moisture supply. The cacti for instance rely on seasonal torrential downpours. In Australia the intermittent occurrence of rain with long drought periods in between probably explains the paucity of these plants. The few species we have like this *Tecticornia verrucosa*, occur on rocky substrates or saltpans where water pools even after very small rains.

Part Two
Ecosystems of the Inland

PHYSICAL ENVIRONMENT

Desert Ranges

Mulga Woodlands

DRIVING FACTORS

EFFECTIVE

All rains produce growth from rock surface run-off. Shaded waterholes provide drought supplies.

Run-off from moderate rains is sufficient to promote growth and recharge groundwater.

SOIL

Usually poor, shallow soils, but compensated by effective rainfall.

Reasonable levels of fertility and good soil moisture storage.

FIRE

Rarely burns, discontinuous fuel.

Moderate frequency, but patchy.

BIOTIC CONSEQUENCES

Critical refugia for most plants and animals. Some food and water is always available.
Most palatable and most nutritious plant foods. Highest productivity and herbivore-carrying capacity. Very disturbed environment (rain, drought, fire, herbivore population bursts), therefore large-scale patchiness in time and space.

Spinifex Grasslands

Chenopod Shrublands

Desert Rivers & Salt Lakes

RAINFALL

Extreme rains needed to flood irrigate these environments.

FERTILITY

Low fertility, sandy soils which dry rapidly.

Excess salts/nutrients. Clay soils yield little water to plants.

REGIME

High frequency, fuel extensive.

Low frequency, fuel limited.

Rarely burns, non-flammable fuels.

Least nutritious plants favour short food chain of termites/lizards. Simple environment.

Despite lower rainfall and poorly drained soils, these nutrient-rich areas have the most palatable plant foods and a high productivity. Most disturbed environments (floods, drought, fire, herbivore population/rabbits), therefore large-scale patchiness in time and space.

Specialised plants and animals only. Low diversity and low populations.

CHAPTER 1

DESERT RANGES

Below, sunk in the desert, lay
the range's crumbling vertebrae
ROLAND ROBINSON, 'Rock-Wallaby'

Give me a harsh land to wring music from,
brown hills, and dust, with dead grass
straw to my bricks.

Give me words that are cutting-harsh
as wattle-bird notes in dusty gums
crying at noon.

Give me a harsh land, a land that
swings, like heart and blood,
from heat to mist.

Give me a land that like my heart
scorches its flowers of spring,
then floods upon its summer ardour.

Give me a land where rain
is rain that would beat high heads low.
Where wind howls at the windows

and patters dust on tin roofs
while it hides the summer sun
in a mud-red shirt.

Give my words sun and rain,
desert and heat and mist,
spring flowers, and dead grass,
blue sea and dusty sky,

song-birds and harsh cries,
strength and austerity
that this land has.
IAN MUDIE, 'This Land'

Oases Locked in Time

'Looking back upon our Expedition a few scenes stand out prominently — the gibber plains at sunset; the bare upland stoney plain . . . the view of the Great Finke Valley where at Crown Point, the river breaks through the Desert Sandstone hills; Chambers Pillar rising solitary amongst the sandhills; the picturesque water-holes of the George Gill Range; the camp, weird and silent, by Lake Amadeus; Ayers Rock glowing bright red in the sunset; the group of graceful palm trees by the side of the rock-pools in Palm Creek and the wonderful gorges amongst the MacDonnell Range.'
(Baldwin Spencer, 1894).

Just as Ellery Creek sketches out the ancient history of the continent, so we can use the Finke to introduce the ever changing ochre landscape of today's arid zone. The Finke cuts a transect through all the typical desert environments of Australia — from the rugged, almost impenetrable flanks of the MacDonnell and James Ranges to the great ribs of the Simpson Desert where it is drowned in its own sand north-west of Lake Eyre. In between, it meanders through grey-green mulga shrubland and sunlit spinifex sandplains; through desolate gibber plains and whitewashed mesas; through salt-saturated lakes and brick-red bright dunes.

HORN SCIENTIFIC EXPEDITION

A torn and compelling land — not high but brutally rugged — made up of domed hills, bevelled ridges and deeply incised gorges and enshrouded in Aboriginal myth attracted the attentions of a party of eminent scientists in 1894, funded by W.A. Horn. The members of the Horn Expedition were looking for a vast mountain oasis, a pristine wonderland of an ancient, 'decadent' flora and fauna, elsewhere vanished, and here locked in time; they believed that the MacDonnell Ranges had existed as an island harbouring ancient life forms when the rest of the continent was 'submerged'.

***A rare daisy,* Helichrysum thomsonii.**

MacDonnell Ranges.

In the middle of a dust-devil desert, their most important scientific findings were a moisture-loving earthworm, 25 species of land snails and 13 species of freshwater snails. The earthworm was, to say the least, an unexpected find. It was found in patches of damp black earth near the sides of waterholes in three locations within the ranges. It appears to be a Gondwanan relict related, albeit distantly, to species found in Queensland, north-west Australia, New Zealand, and scattered localities in New Caledonia, the Falkland Islands, South America and southern Africa. Of the 25 species of land snails, 16 were endemic to the region, their ancestry lost in the mists of time. The 13 freshwater molluscs collected have relatives in South Australia, tropical Queensland and the Northern Territory.

These animals, clearly relicts or 'leftovers' from a past wetter climate, are (still) of great interest not least because, apart from a few reports, virtually no scientific investigations have been carried out on the desert mountains since the Horn Scientific Expedition which, to this day, remains the basic reference for the region.

A Place of Edges

Australia's desert mountains cover 580 000 square kilometres and comprise the Hamersleys in Western Australia and the MacDonnells, James and Musgraves in central Australia. With elevations of up to 1524 metres (Mt Zeil in the MacDonnell Ranges — the highest mountain west of the Eastern Highlands), the ranges rise abruptly from 100 to 1000 metres out of flat desert plains.

"Organ Pipes", Glen Helen Gorge, MacDonnell Ranges.

Most visitors to the desert ranges are struck by the juxtaposition of complex shapes. The place seems to be dominated by edges. There are sheer slabs of steep exposed bedrock rising out of the angular boulders of the footslopes; enclosed pounds or basins; precipitous cliffs and gorges; and twisted and rugged valleys. When it rains, run-off is immediate and rapid, streaming off the cliffs and tearing shallow gullies into the bouldery slopes. The water washes off the flanks of the ranges, funnelling into the fringing outwash slopes. It concentrates in pools and valleys deep within the twisted ranges and seeps from springs on the footslopes. Here, protected from evaporation, which can be as high as 3429 millimetres (at Giles in the Gibson Desert), it remains.

And, indeed, these rare places do harbour oases of plants and animals. Perhaps the most famous of these leftovers is Palm Valley, a tributary of the Finke River in the MacDonnell Ranges.

Palm Valley is named after an extraordinary stand of Cabbage Palms, *Livistona mariae*. Much of the scenery along Palm Creek before reaching the palms is typical of the desert mountains. It is a place of brutal colours dominated by an impossibly blue sky which makes the ochre-red valley walls almost vibrate. Ghostly white gums, *Eucalyptus papuana*, and the deceptive soft-yellow clumps of spinifex cling to pockets of soil in ledges on the walls. But as the valley narrows, hundreds of palms suddenly appear, sprouting incongruously from the rock of the valley floor, their fiery green tops, often 30 metres off the ground, rattling and chattering in the breeze. Pools of water, surrounded by reeds, ripple with fish and water beetles. Cycads, *Macrozamia macdonnellii*, join the spinifex and eucalypts clinging to the desert red rock. It is as if you were stepping back about 50 000 years.

Kings Canyon in the Gorge Gill Ranges of central Australia epitomises the ruggedness of desert ranges. It is an awesome place, dominated by edges, precipitous cliffs, angular boulders and twisted valleys.

Rare places harbour oases of plants and animals. Perhaps the most famous of these "leftovers" is Palm Valley in the MacDonnell Ranges.

The palm is a relict species separated by about 1000 kilometres from the nearest *Livistona* in the north. The closest relative, *Livistona alfredii*, however, occurs on the Fortescue River in the Hamersley Range — a stand similar to Palm Valley's. These palms are all apparently closely related. This probably means that the original population was split recently, geologically speaking.

Whether a plant or animal is relict or not is closely tied up with its ability to disperse. The dragonflies inhabiting the hidden pools of the ranges are not relict because the adults can fly great distances. The fish also, while clogging up shrinking pools during drought, repopulate the region during flood, dispersing with the water. Relict ferns, on the other hand, can only maintain their limited populations around sheltered, permanent springs. The adults and their spores cannot cope with any amount of desiccation.

The last time rivers were running and the central Australian lakes were full was between 50 000 and 30 000 years ago. The central lakes region must have looked very much like Lake Eyre today when it fills with water. No one, however, knows what the environment across the vast tracts of the inland was like. There is some evidence that it possibly did not look much different from the country today after a very good run of seasons. The rivers such as the Finke, when not seasonally running, might have been sprinkled with waterholes up and down their length and their floodplains would have been, at least for part of the year, swampy. It is likely that many relict plants and animals, including the snails and earthworms, were left over from this time. At 50 000 to 30 000 years ago the spores of ferns could easily have been wind-blown or carried by animals from one waterhole to another across much of the continent.

Nevertheless, even at this time, it is difficult to imagine a landscape with continuous tracts of palms. Another agent for dispersal may have been humans. Aboriginal people have lived in central Australia for at least 22 000 years, giving ample time for objects such as the attractive red fruits of the palm to be traded across the continent.

Maggie Springs feeder gully, Uluru.

Within the twisted valleys, water concentrates, protected from evaporation. Murchison Gorge, Western Australia.

The palm, *Livistona mariae*, is restricted to an area of about 60 square kilometres, including minor populations in Finke Gorge itself and a major population in nearby Little Palm Creek. In this area there is a population of about 2000. The permanent water supply bathing the palms' shallow and fibrous root system is provided by the layers of rock through which Palm Valley slices — if the rock could be wrung out, it would drip like a sponge. Because the rocks are gently sloping, the water seeps into the valley which has been dissected in just such a way as to ensure a continous permanent seepage area along a considerable distance.

Another palm-like plant, the cycad, *Macrozamia macdonnellii*, also flourishes in Palm Valley. This plant is found throughout the central ranges on the more shaded western and northern aspects of hill slopes, its roots seeking out the small sumps of water in cliff crevices. The cycad's nearest relatives live in near-coastal areas of the Top End of Australia and in south-western and south-eastern Australia.

A total of 333 plant species have been recorded from this small valley. About 30 of these are considered rare or restricted. Most are restricted to those areas fed by springs and protected from fire.

While Palm Valley, with its green crop of palms, is perhaps the most spectacular refuge, botanically the most important area in central Australia is the George Gill Range which contains the awesome Kings Canyon. This region includes sandstone hills, cliffs and valleys and associated plains, riverine systems, swamps and claypans, and contains almost 600 plant species: about one-third the total arid zone flora. Over 10 per cent of these plants have rare or relict or otherwise unusual distributions. Of the 40 or so rare species, five are known from nowhere else but the Kings Canyon area. There are about 20 relict species and at least eight disjunct plant species, meaning that they have had their populations severed by wind-swept fields of sand dunes.

Livistona mariae *palms at Palm Valley, a remnant population that once flourished when the climate was wetter.*

The cycad, Macrozamia macdonnellii, *is found only in the MacDonnell Ranges complex. Growing in shady aspects, its roots seek out small sumps of water in cliff crevices.*

BULL-DOG ANTS

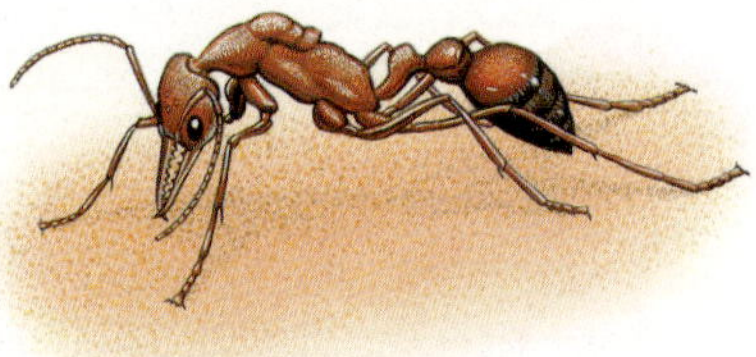

A queen bull-dog ant, Myrmecia gulosa.

There is nothing more delightful than unrolling a swag on a star-clustered desert night amongst the large trees that grow along the flanks of the desert ranges. But beware. This is precisely the favourite habitat of the bull-dog ant, *Myrmecia* spp.

Truly a creature of the underworld, up to 4 centimetres long, agile and aggressive, this ant is always spoiling for a fight. And, like a villain, it will use its powerful sting at the slightest provocation. The bull-dog ant is the most primitive of living ants. The unco-operative workers hunt singly, capturing living insect prey and cutting it up and feeding it directly to the larvae.

The adults, however, have a taste for richer things. They collect nectar from flowers and extrafloral nectaries, which appears to be their main diet when the nest, a sumptuous array of galleries and chambers extending to a depth of 1–2 metres, is without larvae to look after.

Virginity is short-lived in this society. Bull-dog ants fly only on their nuptial flight. On this occasion the females are frenetically seized by a male (and there are hundreds of males to every female). Other males surround the couple until there is a struggling mass of ants forming a ball as large as a fist. After being thus inseminated, the queen excavates a well-formed cell in the soil under a tree or log and starts rearing the first brood of gangster workers.

Islands of Nutrition

Native Pine (Callitris columellaris) *and* Xanthorrhoea australis *eke out water hidden in rocky crevices long after the water reserves in the rest of the landscape have disappeared. Flinders Ranges.*

The concept of a refuge not only applies to the springs and seepage areas hidden within a range, it extends to the entire mountain range itself, even the rocky outcrops. Australia is renowned for its flatness, and it is precisely this flatness which makes the topography so effective; desert uplands funnel and concentrate water.

In the more shaded aspects of desert ranges, such as in the "Lost City" of the George Gill Ranges, trees like this eucalypt can flourish.

With just 15 millimetres of rain, the volume of water concentrating in even minor creeks can be 12 to 40 times more than on the surrounding desert plains. Where water flows, nutrients will follow so that in contrast to the broad sweep of the nutritionally poor landscape the uplands form an archipelago of nutrient richness. These islands are characterised by relatively continuous growth and reproduction.

Because the most predictable thing about rain in arid Australia is its unpredictability, the redistribution of water after rainfall is highly significant for those desert-dwellers which require a predictable supply of water and nutrients. Much of the water is stored like a reservoir in the soils of the creeks and adjacent floodplains where it concentrates after rain. Here it can be tapped by the roots of plants long after the water reserves in the rest of the landscape have disappeared.

As a result, range vegetation is incredibly diverse in both structure and species composition. As might be imagined, vegetation communities occupying a place of edges will be complex, producing a mosaic with abrupt boundaries. For instance, on the cliffs and rock-fall faces where there is little soil development, stunted shrubs with tussock grasses and herbs eke out a precarious existence in crevices; whereas in a more shaded aspect, dwarfed trees such as White Cypress Pine (*Callitris columellaris*) and Ghost Gum (*Eucalyptus papuana*) grow. Further down on the lower slopes where some soil

Shrubs and grasses growing on the foothills of the MacDonnell Ranges.

has developed there are tall shrublands of *Acacia,* mallee, *Eremophila,* spinifex, and the native hop-bush, *Dodonaea.* In addition, in more shaded areas, there can be stands of native fig (*Ficus*) and mint-bush (*Prostanthera*). In the George Gill Range of central Australia even Trigger Plants (*Stylidium*) and Sundews (*Drosera*) can be found in these shady areas. On the scree slopes *Acacia* and *Eremophila* shrubs grow over grasses and herbs such as *Ptilotus obovatus.*

On the low hills flanking the ranges like a rumpled skirt, mallee, *Acacia*, *Eremophila*, Cassia, beefwood (*Grevillea striata*), corkwood (*Hakea* spp.) and white-wood (*Atalaya hemiglauca*), grow over grasses such as *Enneapogon* and *Aristida,* and herbs such as *Helipterum.* In the valleys mixed grasslands occur. On the slopes bluebush (*Maireana*) and saltbush (*Atriplex*) grow over annual and perennial grasses. In the water-courses river red gum, *Eucalyptus camaldulensis,* and *Melaleuca* reach above the grasses.

The life-history strategies adopted by grasses in contrast to herbs in these rich areas are not the same. For example, the grass *Aristida contorta* germinates only after summer rain when temperatures above 23 degrees Celsius have occurred for some weeks. The herb *Helipterum craspedioides*, a common daisy, on the other hand requires winter rain; its seed needs to be saturated at 13–15 degrees Celsius for several days. As a generality, grasses dominate after summer rain and herbs (Asteraceae) after winter rain, though the same species do not germinate and establish each year. This is because, despite the relative stability of these richer areas, the unpredictability of an arid environment allows specific plants to have specific windows of opportunity which only open occasionally. It is a little like a lottery. The right combination of moisture, temperature, nutrients and sunshine will produce the best results for a specific plant.

The flowering and fruiting season of many of the shrubs are also regular, mainly in spring and summer. The regularity is possible because water is relatively predictable in the uplands. But even here, certain plants have certain windows of opportunity: *Acacia victoriae* flowers and fruits in a narrow window in summer, while *A. tetragonophylla* flowers over winter and fruits through summer, and *Eremophila fraseri* flowers and fruits all year round.

The productivity, too, while relatively constant in the uplands compared with the rest of the landscape, will nevertheless vary from year to year. For instance, over a ten year period on Mileura station in Western Australia, *Acacia victoriae* had only one very good year and *A. tetragonophylla* did only slightly better, whereas *Eremophila fraseri* had bumper crops throughout the decade. Not surprisingly most fruit was produced in the watercourses where water is most reliable. Some species, such as *Cassia desolata*, produce fruit in different seasons, depending on where they are in the landscape. In higher country, the fruit sets after summer rain and matures during winter. Lower down in large creeks, however, conditions are too cold in winter for the fruit to mature. Here fruit only matures in spring.

Even on loose scree slopes, such as this one in the Chichester Ranges of Western Australia, enough water concentrates in pockets to allow these eucalypts to grow.

The life-history strategies of animals also reflects these patterns. Because the plants are of better dietary quality, herbivores and birds dominate.

Grasshoppers, for instance, are obvious herbivores in the richer areas. They, as well as many other insects, are at their peak during spring and summer, parallelling the availability of resources. In dry periods their eggs become dormant or go into 'diapause' so that generations of grasshoppers and other insects metaphorically 'hop' through time, landing when there is green growth.

The herbivorous emu, on the other hand, is nomadic. It feeds on fruits, seeds, flowers, insects, and the young growing parts of plants. These are only abundant irregularly so that the animal needs to be constantly on the move, orientating itself towards recent rains. The only time emus are not moving is when the male is incubating.

Emus live in pairs from December to May, maintaining a home range of about 30 square kilometres. Between April and June the female lays a clutch of 9–20 eggs, depending on the amount of summer rainfall. The male then takes over and incubates the eggs for eight weeks, during which time he has the remarkable ability to not eat, drink, or even defecate. He lowers his body temperature by 3–4 degrees Celsius and becomes torpid. When the chicks hatch he shepherds them for five to seven months until independent.

About half of the arid-zone birds are nomadic. These birds are geared to exploit the sudden flushes of food resulting from the patchy and irregular rain that is so much a feature of the Australian deserts. The distances over which birds are able to migrate allow them to treat the arid zone as a system of scattered, temporarily favourable habitats. But perhaps the most surprising aspect of nomadism in the arid zone is that it is not as universal as might be thought, particularly in areas where food is reliable, such as in the ranges. In the Everard Ranges in northern South Australia, for instance, about 60 per cent of the 66 bird species so far recorded are sedentary.

Most are honeyeaters, particularly the White-plumed Honeyeater, Grey-headed Honey-eater, Singing Honeyeater, Spiny-

WITCHETTY GRUBS

***Witchetty Grub of the Goat Moth,* Xyleutes leucomochla.**

Witchetty grubs are the larvae of certain types of cossid moths. They are abundant in patches and feed on the roots of *Acacias* from the outside of the root. The larvae form silken chambers against the root. When the larvae pupate part of the root being eaten is incorporated into the walls of the chambers. Apparently not a great deal of the wood is eaten and larvae may rely on the flow of sap for food, the jaws being used to keep the wounds of the roots active and fresh. The pupae rise from the surface of their burrows in early autumn. After neatly cutting open the silken tops of their burrows and thrusting the lids to one side, the cossid moths then emerge from their pupal shells (which are often found among leaf debris at the base of *Acacias*).

cheeked Honeyeater, Yellow-throated Miner, and the White-fronted Honeyeater. Beyond the ranges honeyeaters have been found to comprise only seven to twenty per cent of the avifauna.

The Owlet Nightjar is widespread across Australia, including arid Australia.

The Spotted Bower Bird, Chlamydera maculata, drinking from the Finke River in the James Ranges.

This high proportion of resident birds is probably due to the stability of food supplies; fruit such as the fig, lerp insects (sap suckers) and ants, for instance, occur year round. These supplies are seasonally augmented by nectar from mistletoe (*Ameyema maidenii*), *Hakea suberea* and by seed arils of *Acacia,* particularly *A. tetragonophylla.*

People usually think of most desert birds as being highly nomadic and moving rapidly to areas where it has just rained and then breeding almost at the smell of it. In fact, while unseasonal breeding does occur, a good proportion of it is unsuccessful and few young survive to maturity. In general, successful breeding in resource rich areas is largely seasonal; most birds respond to the stimulus of increasing day length, and breed in spring and early summer following the cool winters.

More than half of desert-dwelling birds appear not to need free water, in particular carnivorous hawks which rely on juicy prey and most insectivorous species. The groups most dependent on water are parrots, pigeons, grass finches and, strangely, honeyeaters. All but honeyeaters eat dry seeds and these contain the least amount of water of any food. Honeyeaters are an exception, possibly because of their aggressive behaviour (they constantly chase one another when feeding) which builds up a thirst.

The Singing Honeyeater is Australia's most widespread honeyeater. Like many desert-dwellers, the ability to find and conserve water rather than do without it altogether is its most important aspect of desert survival.

These Crested Pigeons are dependent on regular water, and drink at sunrise.

The Black-flanked Rock Wallaby requires a regular supply of nutritious food. It uses the commonsense behavioural adaptation of feeding in the cooler early morning and late evening.

GALL-FORMING BUGS

Adult male **Cystococcus echiniformis** *with baby sisters in tow.*

Galls caused by insects are common on plants. They result from a proliferation of plant cells caused by a chemical produced in the saliva of the insect. Galls occur in all shapes and sizes to suit the insect and it appears that complex molecules produced in the saliva of insects 'tell' the cells of the tree to produce the required shape. Furthermore, specific insects form galls on specific plants.

One fascinating example of the life-cycle of a gall-forming bug reads like something from a science-fiction novel. Females of the bug (*Cystococcus*) cause large woody spherical galls to develop on the stems of their bloodwood eucalypt host. In central Australia these trees grow in the richer areas of the environment, usually around the flanks of the ranges or on the floodouts of creeks. And it is rare to see a bloodwood without these galls, commonly called bloodwood 'apples'.

Once the female *Cystococcus* is established within her gall, she gives birth to a suite of sons. They complete their development within the gall, feeding on a layer of white nutrient-rich material lining the gall cavity. The adult males are fully winged and have a peculiar shaped abdomen which juts out like a thorn or the keel of a ship. When the males are mature, the mother gives birth to daughters. These tiny female maggots scramble onto their brothers and attach themselves to the modified abdomens. The brothers act like buses with the females hanging onto the rails and leave the gall through an orifice which has until then been plugged by the tip of the mother's abdomen.

The adult males are capable of only weak somersaulting flight but they can transport several of their minute flightless sisters to a nearby tree. The females alight, disperse over the foliage and settle down to making galls on young stem tissue. No longer useful, the appendages with which they clung onto their brothers' abdomens disappear and the females become fleshy, featureless birthing machines. Their brothers, once they have carried out the task for which they were born, live less than 48 hours.

Again, overt physiological adaptations to lack of water are uncommon in birds. It seems that the ability to find and conserve water, rather than surviving without it altogether, is the most important thing for survival in the deserts. Most drinking birds water in the early morning or late afternoon. Many of them arrive in large numbers. To avoid congestion, different birds water at different times. Bourke's Parrot comes at or before sunrise, Crested Pigeons, Ringnecks and Mulga Parrots soon after and the Galah and bronzewings not until dusk. The Zebra Finch, Diamond Dove, Spinifex Pigeon, many honeyeaters and bowerbirds are never far from water and drink through the midday hours. Pigeons, finches and honeyeaters have tongues which they can poke out. These birds do not drink by the conventional beak dipping and sipping but by using their tongues as a pump. This makes the process of drinking much quicker at congested waterholes.

The commonsense behavioural adaptations of feeding in the early morning and late afternoon and resting through the middle of the day in the shade are all accentuated in birds. Physiological adaptations do occur, however, and they include lowering of metabolic rates which help to reduce the use of water. The Zebra Finch, Bourke's Parrot, Budgerigar and Spinifex Pigeon all have low metabolic rates.

Strangely enough, nocturnality, widely adopted by other animal groups, is not at all common in birds. Furthermore, the four nocturnal desert-dwelling birds of Australia are not the conventional owls and nightjars (which are found in the arid zone but are not indigenous to it) but the Freckled Duck of the ephemeral inland waters, the Letter-wing Kite of the cracking-clay plains, the Inland Dotterel of the gibber plains and the Night Parrot of the Lake Eyre basin. Of these it seems that only the Night Parrot uses nocturnality to conserve water. The others have evolved to take advantage of specific food items which become available at night.

A desert night is not still. It is full of subtle sounds; the whirr and tinkle of crickets, a stridulating spider, the 'gek-ko' of a gecko (the only lizard that vocalises). Shadows sweep across the star-sprinkled parrot-blue desert sky accompanied by soft high-pitched 'chip . . . chip . . . chip' sounds. These shadows are bats emitting pulses of sound to echo-locate their insect prey.

Bats are mainly an animal of the tropics and subtropics and of the

Left, Crested Pigeons drink by using their tongues as a pump.

Desert ranges have a high proportion of resident birds because of the stability of their food supplies. Flinders Ranges.

62 or so species of bat in Australia, about 15 species roam through the arid zone. Only three bats are restricted to the arid zone; the Inland Brown Bat (*Eptesicus baverstocki*), Little Pied Bat (*Chalinolobus picatus*), and Hills Sheathtail Bat (*Taphozous hilli*).

Not surprisingly, these bats are all insectivores. Their ability to fly makes them very adaptable and one, the Yellow-bellied Sheathtail Bat (*Saccolaimus flaviventris*) has even been found nesting in burrows in the ground and in termite mounds in the Tanami Desert. Bats are most abundant, however, in the desert ranges where they inhabit caves or rocky crannies. They also roost in large trees along the rivers spilling from the ranges. They do not appear to be nomadic or migratory and restrict their foraging activities to resource-rich areas.

Desert-dwelling bats like this Yellow-bellied Sheathtail Bat are efficient exploiters of the abundant nocturnal insects in desert ranges.

BRUCE THOMSON

While this pretty Rock Isotome (Isotome petraea) on Uluru may attract birds, it can cause pain and temporary blindness in humans who pick the flower and then wipe their eyes.

Bats efficiently exploit the nocturnal insect resource by fine-tuning their echo-locating capabilities. In general, high frequencies detect small insects at close range while lower frequencies detect larger insects at greater distances. Many bats coexist by targeting different prey, foraging at different altitudes and at particular times of the night. Unlike other Australian mammals, bats do enter into long periods of torpor, perhaps even approaching complete hibernation, in winter. The desert-dwelling Hills Sheathtail Bat for instance lays down thick fat deposits before winter, implying that it does not feed at all during the cooler months.

Carnivorous marsupials also store fat, not as a thick layer all over their bodies, but at the base of their tails. Fat tails, such as in the Fat-tailed Antechinus (*Pseudantechinus macdonnellensis*) are common in the desert-dwelling insectivores and serve to buffer these animals against food shortage. The Fat-tailed Antechinus, as its species name suggests, is an animal first found in the MacDonnell Ranges. Its general habitat is in the rocky hills of the arid zone where it is a nocturnal hunter, occasionally emerging from the shelter of rocks to sunbathe.

Also feeding in the evenings and into the night are the Euro (*Macropus robustus*) and the Black-flanked Rock Wallaby (*Petrogale lateralis*).

The Australian, or Nankeen, Kestrel is a common raptor in woodlands. It relies on small vertebrates, such as this dragon lizard, for the bulk of its diet. The falcon literally hangs in mid-air minutes at a time, dropping in steps and then finally plummeting onto its prey.

Right, Mulga is highly resistant to moisture stress. Even the silvery blue-green colour of the foliage helps to increase leaf reflectance. Above, mulga woodlands have a characteristic suite of animals, including this handsome Pygmy Mulga Monitor (Varanus gilleni).

DISTRIBUTION OF MULGA WOODLANDS

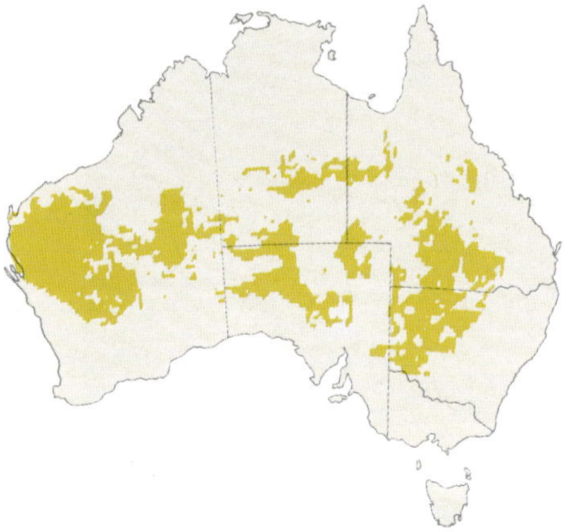

Above, inset, germination of shortlived plants occurs rapidly after rain. Right, termites thrive on the abundant cellulose available in mulga woodlands. Below, the long-lived mulga requires a late summer rain to initiate flowering, followed by a winter rain for the seed to set. Another summer rain is then necessary for germination. This sequence of events may occur only once a decade.

Mulga occurs in many different shapes, so that it is often difficult to identify. This is an uncommon Christmas tree form in the Great Victoria Desert.

While rain is unpredictable in mulga woodlands, when it does rain it can occur in a deluge, causing widespread sheet flooding.

Incredibly, from a rainfall of only 25 millimetres, the equivalent of 140 millimetres of rain can enter the soil in this zone. Water from even small showers can be utilised more readily by mulga for a longer period than, say, the grass and shrub species in the same environment because the water is stored at a greater depth close to the tree.

While mulga can easily cope with the capriciousness of rain in the arid zone, it does not tolerate frequent fire. It has some resprouting ability from the roots but none from the stems. In fact even if only the canopy has been burnt, the tree will die in a few years. While germination from seed is enhanced after fire, it is not a prerequisite. The different abilities to cope with fire between spinifex grasslands and mulga woodlands often results in sharp boundaries separating these communities.

In the outback the word 'mulga' is often used in a general way to lump together any large perennial woody *Acacia* species. For instance, *A. cyperophylla* is known as 'red mulga', *A. calcicola*, as 'shrubby mulga', *A. brachystachya* as 'turpentine mulga' and *A. stowardii* as 'bastard mulga'.

This reflects the incredible amount of variation in phyllode type and habit within mulga (*Acacia aneura*) itself — it is even common practice to enter *A. aneura* several times in identification keys. Mulga phyllode width, for instance, varies from 0.9 millimetres to 12 millimetres, its length from 1 millimetre to 25 millimetres; and the foliage colour from green or grey to silvery-blue. The pods, or legumes, also vary from 1.5 centimetres to 7 centimetres in length and 4 millimetres to 15 millimetres in width. They can be paper thin to leathery, or woody.

There are almost certainly several species in the knot of genetically labile and currently evolving taxa we call *Acacia aneura*, but as yet they have not been teased out by botanists.

In the Great Sandy Desert, mulga occurs in three different shapes; normal, like an umbrella blown inside out; weeping, with softer, greener foliage and drooping

HONEY ANTS

An extraordinary example of adaptation to the desert environment is that of the honey ant, *Camponotus inflatus*. The colonies of these ants open in a series of small ruptures, usually well-hidden by litter at the base of mulga trees, which ramify into a network of tunnels and vaults two metres deep. The all-female workforce forages for honey amongst the mulga, extracting it from extrafloral nectaries and from the honeydew secreted by a red sap-sucking insect which is abundant on the branches of mulga in spring. The workers then return to the nest and regurgitate the nectar into the jaws of 'repletes'; a description which is particularly apt considering their gross bubble-like abdomen. The replete spends her life hanging upside down from the ceiling of a vault by the front and middle legs. During droughts the queen, larvae and workers eat honey held in the living honeypots until the latter become 'depletes' and shrivel and die.

branches; and horizontal, with branches jutting straight out. The different forms grow in different habitats, particularly the horizontal form which grows on rocky areas where the soil moisture regime is unfavourable. Here it grows in stands which are stunted and open. The normal and weeping forms occur together with the latter form favouring flat surfaces with loose ironstone gravel. Mulga itself is not the only dominant woody plant in mulga woodlands. The composition and structure of the vegetation associations reflect their enormous geographical range.

The communities of plants occurring as an understorey to mulga are also varied and complex. For instance, around Alice Springs alone there are 18 distinct groundstorey communities. In continental terms ground-layer species include the grasses *Eragrostis eriopoda, Aristida, Chloris, Dichanthium, Digitaria, Eriachne, Monachather* and *Thyridolepis*. These grasses strongly compete with mulga seedling establishment.

Ephemerals *Ptilotus, Sida, Calotis* and *Helipterum* also occur. About 100 different species of *Eremophila* grow in the mulga lands of Western Australia as well as many *Cassia* and *Dodonaea*. Saltbush and bluebushes (*Atriplex* and *Maireana*) are found in the southern mulga lands.

The cover of trees and shrubs in mulga country is often no more than about 20 per cent but it can be particularly dense, for example where it crops out in groves. In the north of Western Australia and also in the Northern Territory, north of the MacDonnell Ranges, mulga grows in groves which follow the contours of the land. From the air these groves create a pattern like a phenomenal fingerprint. The groves are micro-relief features on the almost flat slopes of the washplains which merge into the sandplains and dunefields of the Tanami Desert.

Slopes as gentle as these (less than 2 degrees) are too gentle for the development of drainage channels, but steep enough to maintain organised patterns of sheetflow. The pattern created consists of a series of contour aligned terraces. In central

After rain, the ground becomes awash. Like a desert tide, the slow-moving water pushes froth into scallops. Seeds, debris and animals are left in strand lines as the water percolates into the red earth.

Australia these are 20–200 metres long and 5–50 metres wide, with alternating wider strips of bare, crusted, or gravelly soil. The soils of the groves are more permeable with a shallow layer of litter sprinkled over the friable surface. The bare slopes of the intergroves act as catchments sloughing water, sediment and seeds which pond in the groves; a remarkable form of natural run-off/run-on irrigation.

On the ground the sight and bubbling sound of one of these sheetflows is unforgettable, particularly since it usually occurs simultaneously with the deluge which caused it. In a startlingly short time the ground is awash; the slowly moving shallow water pushing froth in blood-coloured scallops along the ground. Trapped in this desert tide are seeds, debris and animals which are dumped against trees and left in strand lines as the water percolates into the red earth a short time after the rain ceases.

Above, ephemeral Hill Sunrays (Helipterum saxatile) *flood the ground with colour after rain. Right, other ephemerals such as this* Ptilotus *sp. also occur in mulga woodlands.*

Below at least two different species of ant have made these nests on the ground of a mulga woodland straight after heavy rain. In the background is a ring nest of the Mulga Ant (Polyrachis *sp.*)*. The turrets and chimneys are made by another species of* Polyrachis*. They possibly built these high structures to stop water from flooding their colonies.*

PENNY VAN OOSTERZEE

Kingdom of the Sun

The one thing that is not in short supply in Australia's deserts is sunlight; the desert dwellers have a limitless supply of energy. And structures and materials based on carbohydrates are cheap to make. This means that perennials such as mulga defend themselves against herbivores by incorporating into the leaves indigestible carbon-based compounds such as tannins and lignins; not by growing huge thorns like the prickly* Acacias *in Africa, which use more nutrients than Australia's perennials can afford.

Mulga, however, can afford to be very generous with its cheap carbohydrates; its seeds are packaged with appendages called arils which ants and birds find irresistible; it has specialised glands on the leaves which secrete nectar called extra-floral nectaries; and it can tolerate large populations of scale insects sipping avidly at its sap. The ants and sap-suckers are examples of animals that tolerate the capriciousness of Australia's climate. Like the mulga which grows on through dry times using water stored in the soil, they do not fluctuate in a straightforward way with rainfall; they do not boom or bust.

The concept of boom and bust, in fact, is really a bit of a myth. In reality perennials at least have regular life-history strategies. Mulga, for instance, flowers heavily after spring and summer rain, although it will flower (but not as heavily) at this time even if there is no rain. Only late summer rains will initiate seed production. A winter follow-up rain is necessary for the seed to set. Seeds will be ripe and dropped from the pod the following October and November when they will require more summer rain for germination. Germination will be more rapid if the seed is in an atmosphere rich in carbon dioxide, like that found in decaying litter or in animal droppings. This complicated

Many inland Acacias *are parasitised by mistletoe. Mistletoe fruits are eaten by birds such as the Mistletoe Bird and some honeyeaters. These birds deposit the seed on a branch — where the plant germinates — in their droppings.*

This multi-patterned centipede (Alletheura sp.) *can be found under fallen branches. It is a fierce invertebrate predator.*

Lonely Georgina Gidyea (Acacia georginae) *on the western edge of the Simpson Desert. Australia is recognised as the land of the gum tree but in fact, in the arid zone — which is most of the continent — it is more the land of the wattle.*

window-of-opportunity for mulga — late summer rain, winter follow-up rain and then another summer rain — only occurs about once a decade.

In other countries such as Africa the seeds of many *Acacias* are adapted for dispersal by water and wind or by ruminant wildlife which eat the leathery, nutritive pods. In Australia, where there are no native ruminants, the methods of dispersal are by birds and ants, and Australian *Acacias* have exploited this fact. Like luring a child with a lolly, the *Acacias*'s arils (a dose of cheap carbohydrate) are white or off-white to attract ants which are drawn to white colouration against a black background; their chemical composition may even mimic that of the ants' insect prey. Birds, on the other hand, are particularly attracted to reds and yellows such as the arils on *A. tetragonophylla* and *A. ligulata*.

The plants which use ants to disperse their seed are called 'myrmecochores'. Most Australian *Acacias* fall into this category. Their white or off-white arils have low energy value. Birds are a little more fastidious and the red and yellow arils of the 'ornithochores' are correspondingly higher in energy. Furthermore, the seeds of these are displayed provocatively high up in the branches whereas the pods of mulga are dropped onto the ant-populated ground with the seedpod alluringly open to display.

Not all ants, however, will be attracted to mulga arils. Because they have a low-energy reward they are sought by small bodied ants like *Melophorus* and *Pheidole*. The larger bodied *Rhytidoponera* favour the arils of *A. tetragonophylla* and *A. ligulata* and also the red and yellow fruits of *Rhagodia nutans* and *Enchylaena tomentosa* which often grow under the canopies of mulga trees.

The distribution of these trees and shrubs shows a pattern similar to that of *Rhytidoponera* mounds on which they grow. This strategy takes advantage of the fact that these mound habitats have a greater organic content and moisture holding capacity than the surrounding soils. Ant mounds can have 3 times as much phosphorus and up to 200 times more nitrogen than the surrounding soils. Water also floods deep into the galleries and tunnels of ant nests; a natural reservoir for shrubs and trees which exploit the mounds. Put simply, the high frequency of myrmecochory in Australia may reflect natural selection favouring plants that, in a background of generally poor soil quality, direct their seeds to nutrient-rich micro-habitats.

Mulga trees themselves also act as vegetation micro-environments; there is a tendency for clumping of under-shrubs around relicts of dead mulga and other large shrubs. In fact it is rare for perennial seedlings to become established outside the area directly influenced by prior occupation of perennials. In this sense the mulga trees act as mini-refuges for plants such as grasses and herbs which can survive in these small favourable pockets during drought and recolonise the drier zones, ensuring the survival of the community as a whole when conditions improve.

By using such a strategy these perennial plants avoid the direct impact of extremes. Much of the wildlife of the mulga (and the arid zone in general) uses similar tactics. They burrow, work night shift or generally forage at the least stressful time. The strategy is so successful that water becomes a secondary issue. For the desert dwellers food and water often occur in one packet so that it is food that is most critical. Food is in turn influenced by the availability of nutrients in the system; so that the barren soils on which mulga ekes out a living affect the very foundation of foodwebs dependent on mulga.

After rain, the ground in mulga woodlands will be awash with pretty yellow flowers* (Sida *sp.). The shortlived herbs soon wither and leave these circular prickly fruits, which have the colloquial name of 'teddybear's arseholes'. Below, mulga ants' nests are raised above the ground and thatched with mulga leaves to protect them from flooding.

Below, like miniature radomes, these fungi erupt from the ground after rain in mulga woodlands north of Alice Springs.

Above,* Sturt's Desert Rose *(Gossypium sturtianum) is common in mulga woodlands, particularly along roadsides.
Right, herbivorous animals which require fresh green vegetation, grasshoppers flourish in mulga woodlands after rain. At other times, they are more likely to be found in the richer environments of the flood-outs, or amongst the mixed vegetation on the flanks of the desert ranges.

Many plant-eating animals are seriously constrained in these infertile environments. Except for the ephemeral flush of growth after drought-breaking rains, they have the perennial problem of indigestible food. Grasshoppers will metaphorically hop through time, making sure they are on the scene only during times of ephemeral plenty. But persistence can pay, particularly for those prepared to specialise. Invertebrates and sap-sucking insects, for instance, are found on nearly all perennial plants, including mulga, and remain active for months or even years after rain.

Populations of these groups of consumers are more stable than might be expected. This is because the predictable life-history strategy of plants — regular flowering and seed production — provides a dependable resource on which animals can base their life-cycles.

Social insects, such as ants and termites, represent an extreme development of a persistent life-history strategy and they have done well in Australia's arid zone.

VEGETATION FLUSH

In the outback the wildflowers and green flush of vegetation following drought-breaking rains are particularly spectacular after a long drought. This is because the longer the soil is dry the more nutrients build up, particularly nitrogen. Soil microflora, such as the crusts of lichen and algae that form on the top few centimetres of soil are activated with even a sprinkle of rain spattering the surface. (Roots of plants, on the other hand, require 1–4 days of wet soil to activate.) While active the soil microflora converts atmospheric nitrogen, which is unusable, to mineral nitrogen, an essential nutrient. Soil algae also excrete — often in filaments — enzymes, organic acids, polysaccharides, vitamins and other growth factors. This growing pool of nutrients is drawn on by plant roots following a good rain, resulting in a particularly nutritious vegetation flush if the time between heavy rains has been lengthy. The algae filaments and the stringy nature of the excretions tie soil particles together, improving the soil's water holding capacity and reducing erosion.

The soil microflora themselves have astonishing survival traits. They can, for instance, survive temperatures exceeding 100 degrees Celsius when dormant in the dry soil, and they have the miraculous ability to successfully revive after being dry-stored for at least 83 years.

***Poached Egg Daisy* (Myriocephalus stuartii).**

Ptilotus latifolius.

Empire of Ants

Every time a myrmecologist (ant ecologist) goes out to investigate ants in Australia's arid zone, previously unrecorded species are discovered. There are thousands of species of ants in Australia's arid zone, many are undescribed and unquestionably there are many which have yet to be discovered.

It is often possible to tell the species of animal by the home that it builds. This gash in the red earth is made by a colony of Rhytidoponera *ants.*

In global terms the diversity of ants is related to latitude (and hence temperature) so that, generally, the richest ant fauna is found in the equatorial tropics. New Guinea, for instance, has one of the richest ant faunas with a local fauna of at least 172 species. Australia, however, has double its latitudinal quota. Here you can find an astonishing 100 species in a plot only 20 metres by 40 metres and, in a region at the southern edge of the arid zone, an extraordinary 250 species.

In the arid zone ants can be found nesting in trees, in litter, under logs or stones and in mind-boggling numbers in the soil; in some instances they can be soil specific. They work throughout the day and night, some even zapping about in the middle of summer on soil as hot as 65 degrees Celsius. They forage singly, loosely or in columns streaming across the ground. Ants can be specialists or generalists; seed harvesters, meat eaters, termite eaters even specialised ant eaters which launch army-style offensives into the nests of other species, emerging with larvae and pupae which they either eat or enslave. Ants can be sculptured, brilliantly coloured, iridescent or dull.

Variability in soil is probably a major factor underlying high regional diversity of ants. It is a fact in the arid zone that over a small area soils form a mosaic of types because they have been sorted by wind and water over geological millennia. This in turn results in subtle vegetation changes to which ants respond. There are feedback loops here too, because in burrowing and mixing the soil, generations of ants create particular soil fabrics and even distinctive profiles.

Meat-ants (Iridomyrmex *sp.) cling tenaciously to the stem of an ephemeral plant as floodwaters obliterate their colony.*

If human beings were not so impressed by size alone, they would consider an ant more wonderful than a rhinoceros.
E.O. Wilson

Another important factor influencing the diversity and abundance of ants is the incredible mass of termites in mulga. Termites feed on the feast of cellulose that is available in the mulga lands. They gorge on wood, grass and even the dung of animals. Their cumulative weight reflects the cellulose glut; termites under the ground easily weigh as much as all the kangaroos and cattle on top of it. And one species (*Drepanotermes perniger*) alone of the at least 60 species that occur in mulga harvests 100 kilograms per hectare of dry weight annually. Apart from being important for the development of soils and recycling of nutrients, termites make tasty tidbits for ants.

The native *Iridomyrmex purpureus*, the 10 millimetres iridescent green or blue meat-ant, can be found in huge numbers in the arid zone. They are inquisitive and aggressive and this, combined with their sheer numbers, ensures that they dominate all arthropods on the soil surface. Their tyranny extends to other ants; colonies of *Iridomyrmex* cannot even live with each other so their mutually exclusive colonies form a patchwork on the ground. This sets up patterns in space and time with which other ants must conform.

THE BIG THREE

The prize for the most diverse genus of multi-celled organisms in the world has, so far, gone to the dung beetle (*Onthophagus*), which has about 1500 species. *Acacia* comes close with about 900 species. But the word is out that the big three ant genera, *Iridomyrmex, Camponotus* and *Melophorus*, of which Australia has a monopoly, are also in the running. The problem is that even though scientists working on ants keep turning up new species at a rate that allows them to predict astronomical numbers of ant species, the taxonomy is exceedingly difficult and, for *Iridomyrmex* at least, it has taken 15 years to sort out 10 species — and these were the easy ones.

This Melophorus *ant is a virtual fire-walker, scudding about when ground temperatures are a scalding 65°C.*

subterreanean nest like a plug so that any foraging *Iridomyrmex* searching the soil surface for food and investigating features such as nests will be confronted with a smooth, heavily armoured head and snapping mandibles.

Despite this antisocial behaviour ants are social insects organised into colonies. Typical colonies have one or more wingless egg-laying queens, myriad workers, eggs, larvae, pupae and, at certain times of the year, winged males and females which when triggered by the weather disperse like seeds to establish new colonies. When they land they drop their wings; the male dies and the once-mated female rears the first brood in isolation, or she can be adopted by another colony.

An established ant colony is much like a perennial plant and has an annual cycle. In winter there is not much activity and the colony consists of a wingless queen or queens and workers. There are few or no brood. Activity increases as it warms in spring and early summer. There is an increase in workers as brood production is stepped up. During late summer and autumn the workforce is so large and enough food is brought into the nest that the colony produces winged males and females which are eventually released.

In the meat-ant *Iridomyrmex purpureus*, all workers are the same size and shape but their duties are divided according to age. The youngest remain in the nest cleaning and tending the queen and brood while the older ones venture out. Some of these foragers are trailblazers who search the territory singly, but most work in busy columns. Workers also engage in confrontations, consisting of small aggregations of excited ants, on any common boundaries, even with ants of the same species.

A colony of *Iridomyrmex* is based on one or more mounded nests which are covered with gravel or twigs. Each entrance hole leads to a separate system of galleries and chambers. Although *Iridomyrmex* move from one hole to another during the day they tend to return to the same one at night. One exceptional colony has been found with 87 nests and 1688 entrance holes, but colonies of this size exhaust resources and completely smother any other ant fauna. A colony of this size will have more than one queen and it will increase in stages, expanding into the frontier using nests as outposts which will later be aggregated.

Iridomyrmex do not always have it their own way and a common feature of invertebrates associated with these dominant ants is mimicry. For instance the spider of the genus *Storena* which is a colour mimic, emulates the metallic iridescent sheen of *Iridomyrmex*. It is also a predator of *Iridomyrmex*.

Polyrachis *sp. ring nest.*

LORDS OF THE RING

One of the most elaborate nest structures found in mulga woodland belongs to the ant genus *Polyrachis*. These ants make a variety of striking nests — some with wide cylindrical turrets or cones — with mulga phyllodes placed in a careful fan around the base of the nest. Others form a solid earthen ring covered with, and again incorporating, mulga leaves. The structures are effective against sheet-flooding and the mulga phyllodes stop erosion.

This meat-ant (Iridomyrmex *sp.*) *is found in huge numbers in the arid zone. Inquisitive and aggressive, they dominate all arthropods on the soil surface.*

For instance, the lone forager ant *Melophorus* has been pushed into a time slot when *Iridomyrmex* does not forage — in the middle of the day. *Melophorus* scuds about souped-up on solar power when the ambient temperatures soar into the 40s and the ground can be hotter than 65 degrees Celsius; when even bushflies have called it a day and only the most *Melophorus*-struck myrmecologists are about. *Melophorus* spp. have further partitioned the hot slot; dark species forage when the temperature is around 45 degrees Celsius whereas the red-headed or yellow species are most abundant at 50 degrees.

No one knows what physiological trick allows *Melophorus* to survive such high temperatures but, by *Melophorus* being compatible with *Iridomyrmex* through separation in time, it has led to *Melophorus* exploiting a range of habitats. The genus has radiated into seed harvesters, nectar feeders such as the extraordinary honey ant, specialist termite predators and other predators and scavengers covering almost the entire size range found in ants. One audacious species even occupies the nests of *Iridomyrmex*, carefully secreting itself into empty chambers, emerging onto the surface and foraging in the middle of the day when its unsuspecting hosts are inactive.

Some of the species of another of the great ant genera found in Australia, *Camponotus*, are also tricksters. *Camponotus* usually reduces its interaction with the aggressive *Iridomyrmex* by being a different size to *Iridomyrmex* or by foraging at a different time. *Camponotus* often nests in gaps between *Iridomyrmex* territories. Some, however, mimic *Iridomyrmex*. These species nest within *Iridomyrmex* territory and their smaller-sized soldiers, the minors, are active at the same time of the day as the *Iridomyrmex*. They do not entirely rely on their *Iridomyrmex*-image, however, and are also fast moving so as to avoid encounters with individual *Iridomyrmex*. The major soldiers of *Camponotus* generally remain in their nests near the entrance. Their enlarged heads exactly fit the circular entrance of the

If Australia is the Empire of Ants then* Iridomyrmex *is the emperor. It lords it over all other ants and intriguingly, by squeezing ants into other niches, adds to the complexity of ant diversity.

Pheidole *ants are seed harvesters extraordinaire. The larger ant with the huge head is a soldier and it protects the smaller workers. The seeds are stored in underground granaries which provide a food supply all year through.*

A bull-ant ready to take on the photographer.

Persistence Pays

The Barking Spider (Selenocosmia stirlingi) *is one of the largest and most aggressive of the desert spiders.*

Social insects are particularly abundant in the mulga-land's infertile and uncertain environment. Colonies of ants and termites act as storage organs which can buffer the pulses of production. Because of their flexibility and the stabilising influence of resources provided by perennial plants they never seem to deplete their food supply. Providing old queens are replaced, these colonies are virtually immortal.

A circular hole, as big as a camera lens-cap, is the home of the Barking Spider. The mesh of web around the opening acts like a fishing line while the spider sits and waits for vibrations on the line.

This dominance by invertebrates anchors the food-web which can then ramify into a network of other invertebrate and vertebrate predators.

One of the most stunning invertebrate predators is the Barking Spider *(Selenocosmia stirlingi*). A walk through a dappled grove of mulga north of Alice Springs sooner or later reveals a circular hole at least as large as a camera lens-cap. This is the home of the large (easily covering the palm of the hand), hirsute Barking Spider. At night the Barking Spider can be teased out of its burrow by tickling the mesh of web spread out around the opening with a long piece of tufted grass — a twig will destroy the webbing and give the game away. With luck, the Barking Spider will lunge out of its nest, looking to snatch up a meal, and then drop backwards into the nest.

The Barking Spider is one of the largest and most aggressive of the desert spiders. It is a mygalomorph, that is, a mouse-like spider; as big and as hairy. All the 17 genera of desert trapdoors and funnel-webs, comprising the mygalomorphs are opportunistic feeders. Not surprisingly, many feed almost exclusively on ants and termites. The bigger ones like the Barking Spider supplement their diets with large beetles and even vertebrates. It is one of the few spiders that at least partially roams or forages. The other mygalomorphs generally sit and wait for prey at the mouth of their burrows with the tips of their legs splayed around the rim. When prey passes they lunge and grab and then retreat to enjoy the meal.

Some spiders cleverly extend the territory from which they can gather prey by twig-lining; like fishing these spiders attach litter fragments such as phyllodes of mulga in a fan around the burrow. They then sit and wait for vibrations on the line. The Barking Spiders's mesh of web also functions in this way. Burrows are often aggregrated in the litter zone under the trees where terrestrial insects are most abundant. Spiders which catch their prey beyond the reach of the nest, like the Barking Spider, generally have open-holed burrows in the open spaces where their chase-attack strategy is more efficient. But this exposes even the biggest to predation.

Paradoxically, sheet-flooding is the greatest hazard for desert spiders. The endemic trapdoor group, Aganippi, have bathplug-like doors which sit firmly into a sunken rim. Other spiders have anti-flooding structures which deflect water from open-holed nests.

Female trapdoors live a long time. They take years to reach maturity and can live for at least 20 years. The male, however, wanders for a mate when he is mature, and after mating, dies. Eggs are laid in a silk cocoon in the burrow and the brood is confined to the parent burrow for months, emerging about a year after mating. During this time the female fasts and therefore, because of the need to feed, she cannot breed more than once every two years. Most spiderlings disperse by running out on the ground. A couple of the trapdoor group disperse on the wind, floating on gossamer.

Predictably, to avoid desiccation, mating and dispersion of spiderlings happen during wet weather, after seasonal summer or winter rains, at which time the young can also more easily build their burrows in the wet soil.

Another inhabitant of spider burrows, which usually moves in after eating the owner, is the Striped-faced Dunnart *(Sminthopsis macroura).* This nocturnal insectivore is one of about five dasyurids or carnivorous marsupials that live in mulga. These animals, however, are not restricted to mulga, and their stronghold is in the spinifex grasslands.

Amongst the many lizards that inhabit mulga are the tree-dwelling Gilberts Dragon *(Lophognathus gilberti*), the geckoes, *Rhynchoedura ornata* and *Diplodactylus stenodactylus* and two skinks, *Ctenotus leonhardii* and *C. schomburgkii.* All skinks snack on spiders. This is not surprising considering that, to a lizard, spiders provide a large, juicy parcel. *Ctenotus leonhardii* in particular stuffs itself with spiders, though it also eats abundant quantities of ants, termites and larvae of moths and butterflies when they are available.

One such larvae (of the moth *Chlenias inkata*) appears in spring in large numbers, feeding on the phyllodes of mulga. They have the strangest habit of dropping down on silken threads in their hundreds from the branches at the slightest disturbance, even just a handclap. After an interval as long as 15 minutes they haul themselves up again to their feeding positions. When sated the larvae apparently find their way to the ground to pupate. In captivity these pupae last up to 3 years.

Predictably, such an abundance of insects attracts mainly insectivorous birds to mulga, although nectarivorous birds such as White-fronted Honeyeaters will arrive in numbers during flowering. With the exception of the larger birds, such as the Black-faced Cuckoo-shrike and the Grey-crowned Babbler, the birds of the mulga woodlands are similar to those found in the more fertile woodlands of the desert foothills and floodouts. The Pied Butcher-bird, too, is replaced by the smaller Grey Butcherbird which requires smaller prey and more dense woodland. Birds commonly found in mulga are the Western Gerygone, Slaty-backed and Chestnut-rumped Thornbill, Variegated Fairy-wrens, Chiming Wedgebill (so named because of the sweetly whistled falling chime of four to six notes) and the White-browed Tree-creeper. The Mulga Parrot and Bourkes Parrot can also be found feeding on seeds of the grass understorey.

Fighting a Barking Spider is a matter of life and death for this little carnivorous dunnart* (Sminthopsis youngsoni). *The battle won, however, the Barking Spider makes a juicy and filling meal.

Above, to this Ctenotis leonhardii, *spiders provide a satisfying and thirst-quenching snack. Both the Bearded Dragon* (Pogona vitticeps) *(above right) and Gilberts Dragon* (Lophognathus gilberti) *(right) can be found during the day perched on fallen timber, limbs of trees and termite mounds, from where they forage for insects.*

GRYLLACRIDID

Another invertebrate predator is this ferocious Gryllacridid. The name is derived from 'gryll', cricket-like, and 'acris', grasshopper-like; gryllacridids lie in structure about halfway between grasshoppers and crickets. Based on the quantity of still undescribed material, there may be about 200 species in Australia. In concert with many other invertebrates, lizards and some plants, gryllacridids have their most dramatic evolutionary radiation in arid Australia. They live under the bark of trees such as mulga, burrow in sand and, just as spiders do, often build a cap like a trapdoor. One gryllacridid is known to stopper the top of its nest with a pebble which it pulls into the opening of its burrow and sews firmly closed with silk.

An undescribed species of the genus Ametrus.

A land, as far as the eye can see,
where the waving grasses grow
Or the plains are blackened and burnt and bare,
where the false mirages go
A.B. (Banjo) Paterson, 'The Plains'

A straw-yellow field of waving spinifex (Plectrachne schinzii) provides a stunning contrast to a dark and ominous tropical storm in the Tanami Desert.

Spinifex grows in hemispherical clumps, and the ecosystem is often known as hummock grasslands. More than any other ecosystem, it characterises the deserts of the inland.

CHAPTER 3

SPINIFEX GRASSLANDS

These grasslands are the single most extensive vegetation type in Australia, covering 22 per cent of the continent. In a country known for its infertile soils, spinifex thrives on some of the poorest soils in Australia.

These are the sandplains and wind-sculpted dunefields of the five major Australian deserts; the Great Sandy Desert, Great Victoria Desert, Gibson Desert, Tanami Desert and Simpson Desert, and the skeletal rocky soils of the many arid mountain ranges such as the Hamersley Range, Petermann Ranges, James Range and MacDonnell Ranges — all the areas dominated by spinifex grasslands.

Unlike other grasslands in the world, however, spinifex does not have a complement of large herbivorous animals, such as the bison and pronghorn of the American prairie, or the saiga antelope of Asia, or, of course, the big herbivores of East Africa like the gazelle, wildebeest and zebra. On the contrary, termites are the miniature grazing animals of the spinifex grasslands.

Lizards are, in fact, the hunters of the spinifex grasslands; the Australian answer to tigers, lions and cheetahs. This is not surprising since spinifex grasslands have the richest lizard fauna in the world.

A foodweb dominated by spinifex, termites and lizards is a peculiar thing. And while aridity plays a part, the main reason for it is the incredible infertility of the soil.

Often known as hummock grasslands, this ecosystem more than any other characterises the deserts of the inland. The grasses have a characteristic growth form: a clump of regularly and closely branched long stems from which leaf blades stand out at narrow angles. The effect of this regular pattern of growth in all directions is a hemispherical hummock. The root system of each hummock is diffuse and deep and evenly distributed down to at least three metres. Generally the roots develop from the same nodes as the shoots so that each stem has its own personal water and nutrient supply — an advantage in such a desiccated environment. Being stiff the roots seem to prop up the tussock giving it rigidity.

Triodia basedowii is a 'hard' spinifex — it does not invite any friendly encounters.

During their first dry period, the flat leaf blades become permanently folded. In effect this means that only one side of the leaf is exposed to the drying air, the other side of the leaf being curled in on itself. The leaves are also sclerophyllous, that is they are very hard and fibrous and have silicon granules, called phytoliths, embedded in their epidermis, like microscopic bits of glass. When the leaves become folded, the steel-sharp tips project outwards and strongly discourage intending grazers and passing explorers.

In fact, the word *Plectrachne*, one of the two closely related genera comprising the spinifexes, means 'spear point' in Greek. The other genera is *Triodia*. Both are endemic to Australia. At a continental level the evolution of endemic genera is uncommon in grasses and these genera probably date from the break-up of Gondwana.

Even during the Tertiary period, when rainforest dominated Australia, spinifex, or its progenitors, grew on 'arid' soils — rocky, sandy and often saline. Today about 10 per cent of species grow on saline soils and at least one species has salt glands for the secretion of excess salt. Spinifex species are richest in the refuges of arid and semiarid mountains where the few widespread sandplain and dune species also occur. The rapid climatic oscillations and the general drying of the late Quaternary with the consequent spread of infertile sand sheets and windblown dunes would have favoured the spread of newly formed labile species from within these refuge areas.

There are about 30 different species in the two genera of spinifex. The most common are often categorized as either 'hard' or 'soft', depending, literally, on whether you can grab a handful without puncturing yourself. Most spinifexes have localised distributions, with the desert mountains — the Kimberleys, Hamersleys and MacDonnells — having the greatest number of species.

Only a few species are widespread and they cover large expanses of arid Australia. 'Soft' spinifex *(Triodia pungens)* and Feather-top spinifex *(Plectrachne schinzii)* form most of the communities of the sandplains and some sandridges to the north and west of Alice Springs. 'Hard' spinifex *(Triodia basedowii)* is the most drought resistant species and is common south-west of Alice Springs and in the Simpson Desert and Channel Country of Queensland, as well as in vast areas of Western Australia in the Victoria and Great Sandy Deserts. *T. irritans* is most common in the southern parts of the continent, generally associated with Mallee and also, further north, on the floodplains and along the channels of ephemeral watercourses, and some of the ranges in western Queensland.

Triodia basedowii and a few other less common species have an intriguing growth form that is most visible from the air in localities which have not burnt for a long time. As the hummock grows older and larger it begins to die in the centre and spread outward as a spinifex ring. Kangaroos often camp within the prickly circle, using it as a protective stockade to reduce predation by dingoes and previously the now-extinct marsupial carnivores. In areas which have been unburnt for a long time, the rings may reach diameters of 10–20 metres, at which stage new hummocks appear in them.

Spinifex, of course, is commonly associated with arid conditions. But the term 'aridity' means different things to different plants. For instance, plants adapted to fertile soils will find infertile soils 'arid'. Plants adapted to winter rainfall may not be able to use rain falling predominantly in summer. The plant will suffer drought even though the rainfall amount might not have changed.

The surrounding spinifex grassland is reflected in these water droplets. Note how the top blade of grass has folded to form a 'spear', while the other blades are flat. The blades of spinifex fold after their first dry period.

DISTRIBUTION OF SPINIFEX GRASSLANDS

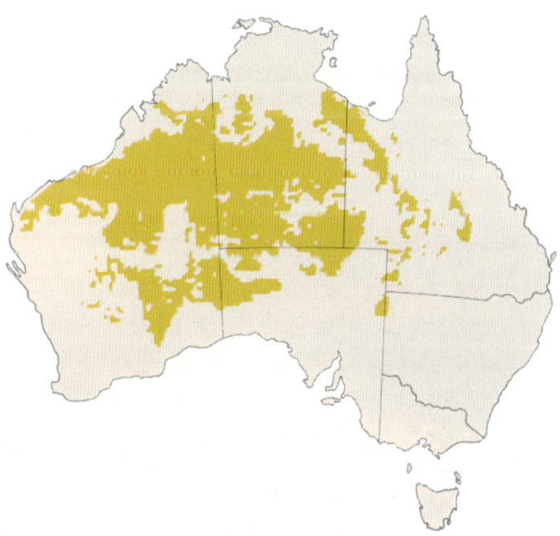

Plectrachne schinzii ***is aptly named Feather-top Spinifex.***

In fact, low rainfall is not necessarily a requirement of spinifex at all since an isolated population on the north coast of New South Wales grows in a 1250–1500 millimetres rainfall zone, on a soil that is unfavourable to the rainforest plants. The adaptations of *Triodia* and *Plectrachne* allow survival under extreme conditions where there is an absence of competitors. Generally however, spinifex flourishes in conditions of low rainfall and infertile soils. It also flourishes under a regime of frequent fire; surviving by committing suicide in a blaze of glory. High flammability due to its dryness and volatile resin in its cells is a feature of spinifex. Like other plant formations, spinifex steadily accumulates biomass after rainfall events but, unlike other plant formations, it does not easily break down and it maintains most of this flammable material throughout its lifetime. In this way it seems to actually encourage fire so that the subsequent passage of wildfire is of unparallelled speed and intensity; a holocaust with convection columns producing enormous, black mushroom clouds.

Fire is a major environmental factor in the modern Australian environment, and those fires in spinifex burn perhaps every five years. Not surprisingly, the trees and shrubs that are commonly scattered across spinifex grasslands have evolved various recovery strategies to cope with fire. Some, like *Grevillea* spp., *Hakea* spp., and the Desert Oak *(Allocasuarina decaisneana)* survive fire because they have a thick, corky bark which protects their living tissue. These perennial plants also regrow from unburnt tissue such as epicormic buds (on stems) and lignotubers (underground).

Others, such as *Acacia* spp. and the short-lived forbs, store seed in the soil. Fire enhances seed germination in these species, restoring populations of individual plants. Others again have seeds which blow into a burnt area from outside. The Desert Poplar, *Codonocarpus cotinifolius*, is a common plant growing in spinifex after fire which recolonises by the wind-dispersal of its seeds.

Some species seem to rise from the ashes. These short-lived fireweeds, such as the appropriately named Fire Bush, *(Cassia pleurocarpa)*, and Kerosene Grass, *(Aristida browniana)*, flourish after fire. One of the edible bush tomatoes, *Solanum centrale*, is also considered a fireweed, in places growing like a crop after fire from seed blown in or carried in by animals.

From a satellite, spinifex grasslands look like artist's brush -strokes. The mosaic of subtle yellows and reds represents the different stages of plant recovery after fire. Different plant groups grow at different times, depending on the amount and timing of local rainfall.

At a regional scale, the amount of plant cover is related to the average rainfall. Further north there is a higher average rainfall. As a consequence, cover is regained more rapidly and fires have a shorter return time. These areas are characterised by plants that can bounce back quickly, either by resprouting or, if recovering from seed, by completing their life-cycles quickly. Examples are resprouting spinifex, mainly *Triodia pungens* and *Plectrachne schinzii* and a high cover of resprouting shrubs. *Acacia adsurgens* and *Acacia lysiphloia* can complete their life-cycles from seeds in just a few years.

Further south, average rainfalls are lower and cover is regained more slowly. The vegetation is characterised by *Triodia basedowii* (a spinifex that recovers from seed), and tree and shrub species that resprout or recover from seed and require a long period between fires to reproduce, such as mulga *(Acacia aneura)*. As a general rule more frequent fire promotes a predominance of resprouters at the expense of long-lived plants which grow from seed.

As* Triodia basedowii *grows older and larger, it begins to die in the centre and spread outward. Kangaroos often camp within the prickly circle.

Spinifex grasslands flourish in a regime of frequent fire. Spinifex biomass does not easily break down, and subsequently, the passage of fire is of unparalleled speed and intensity.

N. PREECE

Topography is another factor which adds to the mosaic-like complexity of the spinifex grasslands. For instance in the dune fields surrounding Uluru, *Thryptomene maisonneuvei* forms pure stands on dune slopes and dune crests in 'old' fire areas. The shrub is fire-sensitive and the clumping affords some protection against fire. On recently burnt areas, however, *Calotis erinacea* and *Helichrysum ambiguum* grow on the dune slopes and crests. On the mid-dune slopes there is a high cover of spinifex in 'old' fire areas, but in the same area shortly after fire *Dicrastylis* spp. and *Keraudrenia* spp. grow.

Some species occur in almost all spinifex areas, having their highest cover in the early post-fire period. *Leptosema chambersii* and *Scaevola parvifolia* are two which, incidentally, produce abundant fruit and nectar soon after fire. *Aristida holanthera* and *Rulingia loxophylla* similarly have a high cover after a fire.

As a general rule, while perennials (such as spinifex) occupy a limited area in this early post-fire period, short-lived grasses and forbs fill the space. As perennials dominate, short-lived species disappear and there is a slow but certain dominance by scattered woody perennials and abundant spinifex.

In October 1872 Ernest Giles, the first European into the Gibson Desert was having a hard time of it. '. . . the natives were about, burning, burning, ever burning; one would think they were of the fabled salamander race, and lived on fire instead of water.'

In a way, he was right because with the arrival of the first Australians into the heart of Australia at least 22 000 years ago, came firestick farming. Aborigines used fire for hunting, regeneration of food plants (a majority of their staple plant foods was from fire regeneration plants), signalling, warmth, play and to clear paths through spinifex. The result was to produce a mosaic of patches of country at different stages of recovery from fire — it provided habitat for plants and animals which required differing stages of regeneration and it eliminated extensive wildfires by providing low-fuel areas as firebreaks.

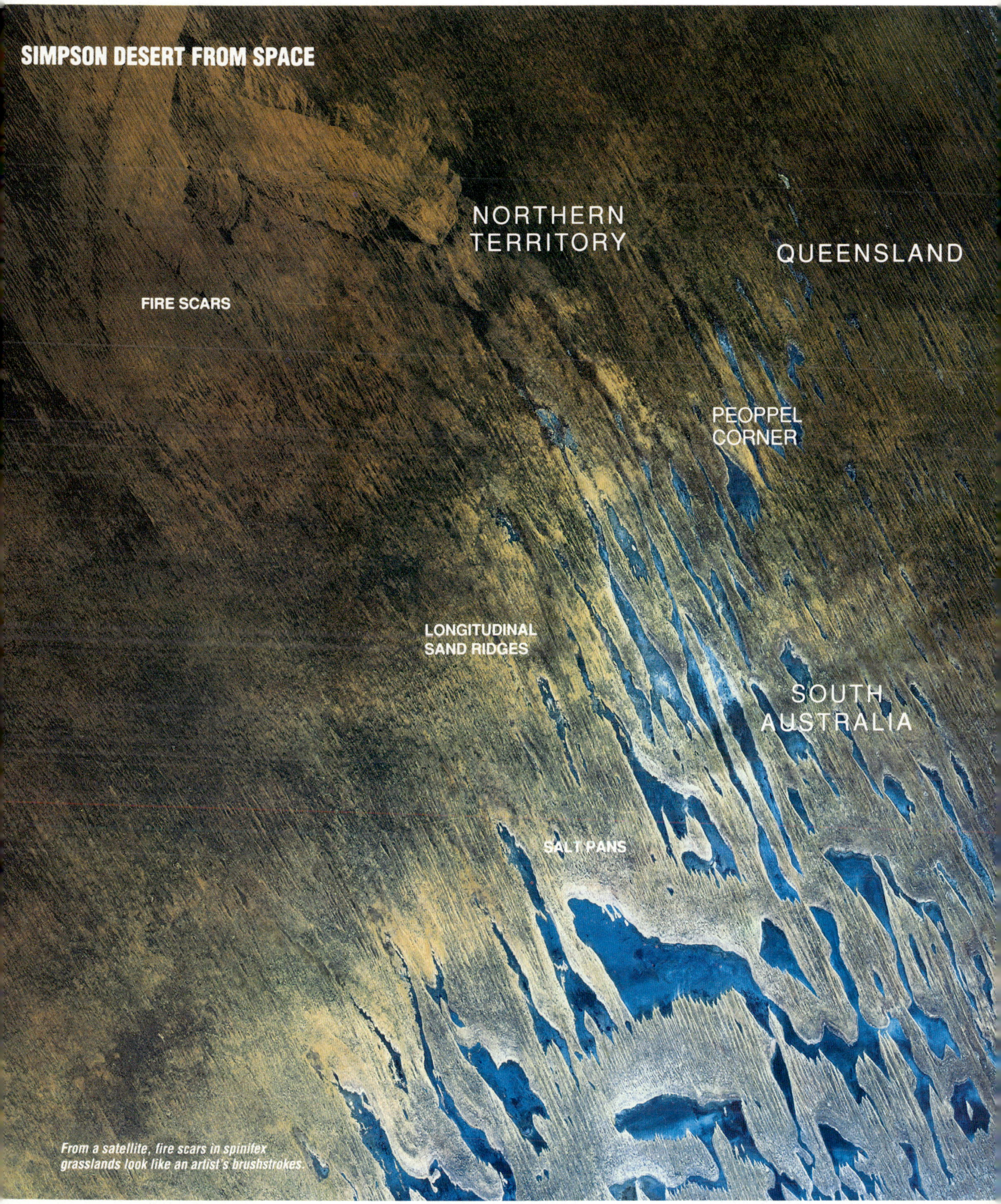

From a satellite, fire scars in spinifex grasslands look like an artist's brushstrokes.

Fire and Desert Mammals

With the movement of Aboriginal people into settlements in the 1930s and 1940s wildfires returned to central Australia. Particularly after wetter periods, such as occurred in 1973/74, fires razed tracts of land the size of European countries: wildfires in the summer of 1974/75 burnt 120 million hectares of land.

Fire in spinifex grasslands: a holocaust with convection columns producing enormous black mushroom clouds.

Because fire is a major environmental factor in spinifex grasslands, plants that grow here must be able to bounce back quickly, either by recovering quickly from seed or by resprouting.

A number of desert mammals were known to be widespread at the time Aboriginal people were moved into settlements, but by the time the first modern detailed investigations on desert mammals were carried out, as late as the early 1970s, many of these such as the Burrowing Bettong *(Bettongia lesueur),* the Golden Bandicoot, *(Isoodon auratus*), and the Desert Bandicoot *(Perameles eremiana*) had vanished. The considerable change in the fire regime has been closely implicated in their demise.

One small macropod associated with the great spinifex deserts, the Rufous Hare-wallaby *(Lagorchestes hirsutus*), while still extant, is critically endangered. About the size of a European hare, this shaggy, rich sandy-coloured animal was once one of the most abundant and widespread macropods. Now the species is known only from the Bernier and Dorre islands, off Western Australia, and from a small region of the Tanami Desert. This mainland subspecies, concentrated around a palaeodrainage system, is morphologically and behaviourally distinct from the island populations. Current estimates put its population at as low as 20 individuals and still declining.

The major collapse of this population 40 to 50 years ago coincided with the movement of Aboriginal people into settlements.

In the Tanami Desert, the Rufous Hare-wallaby lives in dense patches of the spinifex, *Triodia pungens*, which it uses primarily for shelter. It uses an adjacent area with less spinifex for feeding. As conditions become drier the animal moves still further afield into areas dominated by the succulent shrubs within the paelodrainage system.

The diet of the animals is highly variable but predominantly herbivorous. It prefers seeds and fruits when available. Leaf and stem material from grasses are major food and, when all else fails, spinifex itself becomes a food source, supplemented by insects.

This ability to recognise change within the environment and to respond to it is an important adaptation of mammals to life in the spinifex grasslands where productivity is extremely variable. The Rufous Hare-wallaby needs a mosaic of vegetation structure and diversity and small-scale patchy fire is clearly implicated as an important force in creating suitable habitat.

Spinifex 'likes' change because change means that competitors lose their footing. Thus, the rapid climatic oscillations and general drying of the Late Quaternary with the consequent spread of infertile sand sheets and windblown dunes, where little else could survive, was serendipitous for spinifex.

N. PREECE

ROLE-CALL OF THE DEAD AND DYING DESERT-DWELLERS

Endangered Extinct

Pig-footed Bandicoot *Chaeropus ecaudatus*
Desert Bandicoot *Perameles eremiana*
Western-barred Bandicoot *Perameles bougainville*
Golden Bandicoot *Isoodon auratus*
Sandhill Dunnart *Sminthopsis psammophila*
Long-tailed Dunnart *Sminthopsis longicaudata*
Kowari *Dasyuroides byrnei*
Bilby *Macrotis lagotis*
Lesser Bilby *Macrotis leucura*
Numbat *Myrmecobius fasciatus*
Western Quoll *Dasyurus geoffroii*
Red-tailed Phascogale *Phascogale calura*
Gould's Mouse *Pseudomys gouldii*
Pebble-mound Mouse *Pseudomys chapmani*
Alice Springs Mouse *Pseudomys fieldi*
Desert Mouse *Pseudomys desertor*
Dusky Hopping-mouse *Notomys fuscus*
Short-tailed Hopping-mouse *Notomys amplus*
Long-tailed Hopping-mouse *Notomys longicaudatus*
Greater Stick-nest Rat *Leporillus conditor*
Lesser Stick-nest Rat *Leporillus apicalis*
Central Rock-rat *Zyzomys pedunculatus*
Desert Rat-kangaroo *Caloprymnus campestris*

Termites

Being virtually unpalatable to herbivores, this huge volume of leftovers is a bonanza for termites. This is immediately obvious on a visit to the Tanami Desert north-west of Alice Springs. Here are 800 termite mounds per hectare, and foraging tunnels run to virtually every single spinifex hummock, opening from a vertical offshoot onto the surface beneath the centre of the hummock.

When the time is right, termite alates escape in clouds from small holes in the ground.

Each of the termites' castellated cities contain tens of thousands, sometimes millions, of termites. Rambling subterranean galleries, with ceilings like cement pavements, also abundant as are nests in wood; low trees and shrubs, while sparse, are also a feature of hummock grasslands.

Termite fauna is also surprisingly rich in variety. In a small area near Alice Springs, for example, up to 50 species of termites can be found. They recognise different spaces or 'niches' within one ecosystem. These niches are multi-dimensional, not just geographical. The different species of termites, for instance, manage to co-exist by partitioning the food they eat, the time they gather it and the timing of the annual nuptial flight of winged adults, or alates, so as to avoid conflicts of interest.

The alates are the young king and queen termites which set out to establish new colonies. These royals are the only termites in the colonies of millions which grow lacy wings. Their flight from the nest is prompted by several days of rain in the right season. Then they escape in clouds from small holes in the ground. This is a rarely observed and most extraordinary sight and when it occurs at night a torch will act as a beacon drawing literally thousands of alates which have no respect for eyes, ears and any opening in clothing. This lover's flight is synchronised with other colonies of the same species. The flights are usually short, and after landing the termites unhinge their wings, pair off, and the couples dig into the soil and establish new nests.

The Tanami Desert is virtually pincushioned with termite mounds. The biomass of termites under the ground is probably greater than all the other animals living on the surface.

It had been raining heavily for days. Even the extremely permeable desert sands were soggy. It was hot and humid that night in the Tanami Desert. A flash round with the torch revealed this incredible sight, rarely captured on film: boiling from small holes in the ground came the young king and queen termites on their nuptial flight. All over the sand plain, the alates were leaving in a synchronous flight.

Apart from the nuptial flight, termites are completely sedentary, their movements being confined to their nest systems. Termite colonies can in fact be likened to perennial plants. Like roots, termites collect water from the soil, tunnelling to stupendous depths, perhaps as much as 100 metres. Their energy is gathered above ground, like a plant, and stored in the galleries. Even the flight of the alates can be seen as a dispersal mechanism, like winged seeds.

Termites are much more abundant in the dry regions of Australia than in the moister regions to the south. As a small animal in the desert, they have the serious problem of desiccation to solve. The following example emphasises the point. Say an 80 kg person shrunk to the size equivalent of an 8 mg termite, your surface area to volume ratio would increase dramatically, around 200 times. This would result in you losing some 200 times more water than when you were normal-sized.

Termites initially get around this problem by increasing the humidity of their mounds. Walls are built thick to provide good insulation. Then water is brought up from deeper soil. Termites can even take stored grass into damp areas and use it as a sponge. They also manufacture their own water metabolically. The combination of water from the soil and metabolic water results in most termite mounds having relative humidities within the galleries of close to 95 per cent.

One termite builds mounds which act literally as a still. *Coptotermes brunneus,* found in the north-west of Australia, builds thick, impermeable clay mounds two metres high, the upper parts of which have bulbous vase-shaped galleries. These galleries are connected by vertical tunnels to a large underground live-in chamber of uniform temperature and humidity. Metabolic water from this chamber and water from the soil saturates the air spaces throughout the whole nest, including the upper part of the mound which is more susceptible to temperature change. At night the temperature drops an average of 12 degrees Celsius and the water vapour at the top of the mound condenses and drains into the base of the vase-shaped galleries from which the termites can retrieve it.

Adaptations such as these, combined with the basic organisation of termites has meant that, for these animals, lack of water has become relatively unimportant for survival in Australia's deserts.

FLORA OF THE SIMPSON DESERT

Poached Egg Daisy.

The vegetation of the Simpson desert forms distinct zones corresponding to crests, slopes and swales or interdunal corridors. Canegrass *(Zygochloa paradoxa*) is one of the few plants that can make a living out of the coarse highly mobile crests. The spinifex *Triodia basedowii*, grows on the stable slopes and in the sandy interdune corridors. A range of shrubs such as *Acacia, Eremophila* and *Grevillea*, and seasonally abundant herbs are also associated with these slopes and corridors.

The less sandy corridors show a wide range of soil type, such as old alluvial flood-outs, saltpans and gibber, indicating the extent to which the sand has blown back across the environment. This variety is reflected in the vegetation which ranges from low open woodlands or tall, open shrublands of *Eucalyptus microtheca, Acacia georginae, A. aneura, A. kempeana* and *Hakea,* to low open shrubland of *Atriplex vesicaria, A. nummularia, Maireana aphylla, Holosarcia spp.* and *Muehlenbeckia cunninghamii*, to sparse short grasslands with scattered low trees and shrubs.

There are some 800 species of plants documented for the region, most from dunefields with their varied habitats, and floodplains which are refuge areas bringing in water from outside.

Far from a barren dune field of wind-swept sand the Simpson is a diverse region supporting large populations of mammals and birds and lizards typical of the spinifex grasslands, including one animal endemic to the region the Eyrean Grasswren.

The aptly named Poached Egg Daisy (Myriocephalus stuartii).

Left, the long tap root gives this plant's common name of Wild Parnip (Trachymene glaucifolia).

Below, Ptilotus latifolius.

Above, Scaevola collaris.

Left, Cunningham's Bird Flower (Crotalaria cunninghamii).

*Like tiny pink stars, this Desert Fringe Myrtle (*Calytrix longiflora*) is a decorative plant of sand dunes.*

ARID ZONE FLOWERS

The 'Dead Heart' and the 'Red Centre' have often been used to describe central Australia. Nothing could be further from the truth. Particularly after winter rains, central Australia becomes a floral centre, with large areas awash with yellow, pink, white and purple ephemerals.

Left, Brunonia australis.

Below, Sturt's Desert Pea (Clianthus formosus)

Right, Forrest's Featherflower (Verticordia forrestii).

Parakeelya (Calandrinia polyandra*)*

Right, Ptilotus exaltatus *in a field of yellow daisies.*

Eremophila abietina.

Left, Brachychiton paradoxum.

Above, the Upside-down Plant (Leptosema chambersii)*, so named because the flowers are at the bottom.*

Birds

Birds, like mammals, need a constant supply of food. Therefore, bird populations shift rapidly in response to changing conditions. Nomadic birds quickly locate areas where rain has generated a flush of plant growth and insect activity.

Under these conditions in the monotonous spinifex sandsheets of the Tanami Desert, where no surface water exists, as many as 50 species of birds congregate. These include flocks of Budgerigars and Zebra Finches as well as explosive numbers of Crimson Chats and a variety of honeyeaters such as the Black, Grey-headed and Pied Honeyeaters.

As conditions deteriorate the nomadic birds move out leaving only a few sedentary birds which can eke out an existence on the insects, occasional seeds and flowers. Examples are Brown Falcons, Singing Honeyeaters, Black-faced Woodswallows, Variegated Wrens and White-winged Wrens.

Rarely are birds specific to the spinifex grasslands. The Striated Grasswren, Eyrean Grasswren (restricted to the Simpson Desert) and the Rufous-crowned Emu-wren are exceptions. And these are patchily distributed and hard to find. The Eyrean Grasswren for instance was first collected in 1874–75 but then not positively identified for a century until it was collected again at Peoppel's Corner in the Simpson Desert in 1977.

The Eyrean Grasswren lives amongst the hummocky clumps of dune cane grass, *Zygochloa paradoxa*, and spinifex that clothe the treeless sides and crests of the sea of Simpson Desert dunes. As an adaptation to their conditions, their external openings are small to protect them from windblown sand. They seem to make the most of tiny patches of productive habitat, husking and crushing the large seeds of the dune cane grass with their bills which, unlike other grasswrens, are quite finch-like. They also eat insects, particularly ants and bugs.

The most widespread of the grasswren species is the Striated Grasswren which occurs patchily over the whole arid zone, except the Simpson Desert (occupied by its close relative the Eyrean Grasswren). They are true birds of the spinifex — whether on sand plains, dunes or stony hills — hopping and bounding about like rubber balls in search of their insect prey.

Striated Grasswrens have an extraordinary colour grade across the continent; from the fiery russet back and creamy breast of the shimmering ironstone ranges of the Pilbara through to the less intense red back and immaculate white breast of the Tanami and the Great Sandy Desert, and finally to the grey-brown streaked bird of the south-east. In arid areas natural selection favours red-brown pigments in small ground-dwelling birds. These blend with the terrain and hide the birds from predators. As the desert sands become redder north-westwards, so the grasswrens become progressively rustier.

Amongst the tiniest of birds in Australia are the Rufous-crowned Emu-wrens. As the name suggests the crown of the bird is a rich red and in the male the face and under the chin is a deep sky-blue. These birds are also patchily distributed in the spinifex grasslands, particularly where dense tangles of spinifex occur.

The reason for the patchy distribution of the grasswrens and emu-wrens is unclear, but in the Tanami they seem to be concentrated around the palaeo-river systems where more nutrients produce bigger spinifex clumps and probably greater food resources.

The Variegated Fairy-wren is the most widespread of Fairy-wrens. Occurring right across the continent, in arid areas it can be found in bushes along ephemeral watercourses.

The beautiful Pink Cockatoo is generally scarce. It needs to drink regularly and favours the seeds of the Native Pine (*Callistris*) and *Acacias*.

Most people associate Kingfishers with water but the Red-backed Kingfisher (*Halycon pyrrhopygia*) can live far from water and ranges across Australia's desert regions. They are frequently found along the River Red Gum-lines rivers of sand which streak the spinifex grasslands.

Lizards

If there is one animal that is universally successful in arid areas it is the lizard. This success reaches its peak in Australia where there has been a virtual explosion of lizards, the trigger being a web of forces rooted in the infertility of the soil and the unpredictability of the climate.

'Thorns' of the Moloch horridus.

The Sand Goanna (Varanus gouldii) *grows up to 1.6 metres.*

Lizards are relatively diverse in most arid areas of the globe because it 'costs' less to be cold-blooded and not have to maintain a constant temperature in areas of extreme heat and cold. Lizards can also become inactive, virtually switching-off, during periods of stress. During the dry periods, resources dwindle and food becomes scarce. Most mammals — an important exception being the insectivorous mammals — are unable to make a living in these conditions.

The immediate consequence of this is that there are few mammalian predators. In a system dominated by reptiles, reptiles themselves become the most effective predators. The varanids are the top predators in the spinifex grasslands. The largest varanid, *Varanus giganteus*, grows up to two metres long and is the second largest lizard in the world. Even the smallest pygmy goannas are predators and one, *V. caudolineatus*, may deliberately harvest geckos' tails (the incidence of broken geckos' tails being up to 70 per cent), the geckos themselves being too large to fit into their mouths.

Indeed, arid Australia could easily be called The Land of the Lizard. In the spinifex grasslands just south-west of Alice Springs, there can be up to 440 lizards per hectare and up to 42 different species can be found living together. This coexistence of lizards has no ecological parallel anywhere else in the world.

Some lizards burrow, some climb, some like cover, some like the wide open spaces and they all work in shifts around the clock. The structure of the spinifex grasslands is one factor that lends itself to this fine-scale partitioning. For instance, in the Great Victoria Desert in Western Australia, an undulating sandy region with occasional long sandridges, rocky breakaways and salt lakes, as many as nine different geckos can be found living closely together. They manage to coexist by partitioning space and food. Some geckos, like *Gehyra variegata*, are arboreal, or tree-climbers, others are found within spinifex clumps, and others, like *Diplodactylus ciliaris*, forage only in the open spaces. There are also differences between the size and types of prey they eat so that geckos with larger heads, like the Knob-tailed Gecko *(Nephrurus levis)*, eat larger prey than those with smaller heads, like the Beaked Gecko *(Rhynchoedura ornata)*. The Knob-tailed Gecko, in fact, eats the little Beaked Gecko. The Beaked Gecko and two others are termite specialists and, revealingly, there are little detectable differences in how these animals partition their habitats, presumably because there is no need, there being super abundance of termites. *Rhynchoedura ornata* and *Diplodactylus conspicillatus* even have the same resting places — down (unused) spider burrows.

Geckos form the bulk of Australia's abundant nocturnal lizards, and while they partition food and space, they all seem to come on nightshift together.

The little 'horn' above each eye adds a touch of menace to the Spiny-tailed Gecko (Diplodactylus ciliaris).

TRIODIA BASEDOWII

Most spinifexes are resprouters and *Triodia basedowii*, a seed recruiter, is an enigma. Not so much because it recruits by seed but more because, while flowering is copious, less than 6 per cent of flowers produce a viable seed. Given the scarce resources available to it in terms of water and nutrients, this is a most surprising result. What is behind this apparent flagrant waste of resources remains a mystery.

Having just shed its skin, this Knob-tailed Gecko (Nephrurus levis) *is looking fit to kill (top). In fact, the Knob-tailed Gecko feasts on smaller geckos. Beware its satanic smile (above).*

The Blind Snake (Ramphotyphlops australis).

The Pygmy Goanna (Varanus eremius) grows to about 40 centimetres and preys on skinks.

THE LITTLE BLOTTER

The bizarre looking Thorny Devil *(Moloch horridus)* has an unusual way of absorbing water. Separating the scales of the lizard's skin are narrow grooves which form a continuous network to the mouth. If the Devil is placed in a puddle, water runs up the legs and then spreads over the surface of the trunk by capillary action, eventually reaching the mouth. But this hardly seems an evolutionary advance, seeing that any pool from which water could be pulled across the surface of the skin would be deep enough to drink directly. It is in fact the grotesque spines, long thought to be only a defense mechanism, which act as a focus for condensation of droplets of vapour or dew which then make their way by capillary action to the mouth.

The most spectacular use of all the dimensions of a habitat occurs in the skink *Ctenotus*. And, not surprisingly, spinifex has been intimately involved with the evolution of this genera. *Ctenotus* are slender and secretive lizards apparently adapted to 'swimming' through spinifex and because they flee into spinifex clumps at the slightest disturbance, they are rarely seen by humans. Two goannas in particular *V. eremius* and *V. gouldii*, prey on *Ctenotus*, as do the elapid snakes, making them a wary and nervous creature.

Thirty-eight of the 78 species of *Ctenotus* occur in arid Australia. But only in the desert dwellers do up to 11 species coexist, rather tightly packed ecologically speaking. Some forage between plants in the open space — as a rule these have evolved longer hind legs to make escape from predators more effective. Others forage within the tussocks. While they all eat large amounts of termites, they are opportunistic feeders taking whatever prey they encounter. Differences in body sizes mean that large lizards concentrate on large prey, and small lizards on small items. Those with higher body temperatures are active in the middle of the day and those with lower body temperatures are active when the temperatures are cooler, some even becoming most active in winter. The Panther Skink *(Ctenotus pantherinus)*, an exquisitely patterned lizard with light coloured spots with dark borders, is one of these 'cool' lizards. It is a termite specialist and is also active at night.

The ecological co-existence of so many similar animals, then, is done by clever manipulation of body size, use of habitat and micro-habitat, and clocking on and off at different times. In other words lizards are so 'well-made' for the spinifex ecosystem that they can virtually be 'packed-in'. In addition, a surprising new finding suggests that in some circumstances *Ctenotus* actually go underground for a very long time — maybe even for a year.

The Central Australian Blue-tongued Lizard (Tiliqua multifasciata) *is in fact a very large skink.*

Beautifully patterned gecko, Heteronotia binoei.

The exquisitely patterned Panther Skink (Ctenotus pantherinus).

HAL COGGER

LERISTA

One of the physiological adaptations animals can undergo when adapting to life in desert sands is reduction in size of limbs or even complete loss. In this way animals can more easily move, or more accurately swim, through the sand.

Skinks are renowned for the frequency with which they have undergone limb reduction and loss. Of the non-skinks — about three-quarters of all lizards — limb loss has occurred about 3 or 4 times. Skinks on the other hand, comprising only a quarter of all lizards, have lost their limbs around 13 times. The best example of limb reduction, not only in lizards but in all vertebrates, occurs in the skink *Lerista*.

Up until very recently, *Lerista* were considered an uncommon lizard. With the advent of pitfall trapping, where arrays of buckets are sunk into the ground and connected with small drift fences, the real abundance of this secretive lizard has been revealed. In the Tanami Desert, north of Alice Springs, for example, *Lerista* individuals comprise 62 per cent of all lizards. After the skink *Ctenotus*, *Lerista* is the second largest group of lizards in Australia, with 51 species so far.

Limb reduction in *Lerista* is an obvious adaptation to moving through sand. It is of great interest to scientists because, in *Lerista,* it is one of the most graphic demonstrations of a possible course of evolution. *Lerista* is a textbook example showing all the stages of limb reduction, starting from the bones in the toes. Incredibly, *Lerista* gains vertebrae at the same rate as it loses toes, thereby augmenting its sand-swimming motion at the expense of its limbed locomotion.

Dasyurids

A mammal's physiology, on the other hand, is not up to this 'packed-in' use of desert environments. Mammals are energetically 'expensive'. For a start, the small desert mammals only become inactive (and use torpor) under extreme stress and then only for a day or two. They normally require relatively large quantities of nutritionally rich food every day.

Dasyurids like this dunnart (Sminthopsis crassicaudata) *store fat around their tails.*

Carnivorous marsupials such as this dunnart (Sminthopsis youngsoni) *rarely need to drink. This Barking Spider's diet contains at least 60 per cent water.*

The widely held belief that Australian arid-zone mammals are depauperate, however, is not true. Research using subfossil remains from the surface of cave deposits, deposited by owls which accumulate remains of ground-dwelling small mammals, shows that species richness of native rodents and marsupials has declined in the arid zone by a staggering 44 per cent and 41 per cent respectively of pre-European numbers. It is only now being realised that the arid zone contained more desert dwellers

than anyone had imagined and that the ecological characteristics of the communities were different from those of today, though now these are difficult to determine.

And still no other arid zone contains the richness of Australia's insectivorous small mammals, the carnivorous marsupials. Not surprisingly, about half of these arid-zone insectivorous marsupials occur in the infertile spinifex grasslands where a foodweb built on insects (termites) has ramified into a wider web of invertebrates.

Nine species of these Dasyurids — dunnart, *Ningaui, Antechinus* and Mulgara — occur in the hummock grasslands, and all but one, an *Antechinus*, are confined to it.

Desert-dwelling Dasyurids rarely, if ever, drink. This is because their food supply, small vertebrates and invertebrates, contains at least 60 per cent water, so that food and water come in one juicy parcel. They also avoid heat stress by being nocturnal and, during the day, resting in a burrow.

While most of their adaptations are behavioural, Dasyurids are unusual because they store fat around their tails. This is obviously a physiological safeguard which buffers the animals against food shortage in an area of low, unpredictable rainfall and resources which fluctuate wildly. Nevertheless, insects, perhaps more than other deserts elsewhere, are active year round. Given this comparatively reliable resource, it is not surprising that the carnivorous marsupials are seasonal breeders rather than opportunists, like the rodents.

Spinifex Hopping-mouse

But when there is a prolonged superabundance of resources, rodents, such as the Spinifex Hopping-mouse (Notomys alexis)*, will irrupt, thousands of individuals suddenly invading previously uninhabited areas and then, just as quickly disappearing. The Spinifex Hopping-mouse, a pretty sandy-brown mouse with white belly and long tufted tail, is an opportunistic breeder, probably responding to changes in vegetation after a period of rain.*

Young Spinifex Hopping-mice in a nest of dried plant material. Desert-dwelling mammals avoid temperature extremes by creating 'air-conditioned' homes.

Spinifex Hopping-mouse (Notomys alexis).

Like most other mammals of the desert, the Spinifex Hopping-mouse does not need to drink. Moreover, it will actually gain weight on a diet of dry seed. In fact, this intriguing Australian rodent has the distinction of having the most concentrated urine ever recorded for mammals, the most efficient kidneys of all the mammals on Earth. The Spinifex Hopping-mouse has another neat physiological trick. Occasionally during the summer, temperatures in its burrow will be higher than the animal's temperature of 38 degrees Celcius. Humidities in burrows are close to saturation, so that cooling by evaporation (sweating) is not possible. In these conditions the animal is able to become hyperthermic, raising its body temperature above the ambient temperature so that, to it, the ambient temperature feels relatively 'cool'.

While it is easy to be fascinated by these sorts of adaptations, it is important to remember that they are extreme; only one end of a spectrum of incredible life-history strategies. Indeed most of Australia's desert-living animals have behavioural rather than physiological adaptations. For instance, they build burrows to create their own micro-climate — just as we protect ourselves by living in shelters (houses), in effect creating our own micro-climate. Also by feeding and being active at night, the desert dwellers choose the most comfortable conditions in which to be active. Even the Spinifex Hopping-mouse with its impressive physiological repertoire has to be nocturnally active and diurnally fossorial (living underground) to allow it to be totally independent of external water. The same applies to other animal groups such as invertebrates.

GRANIVORY AND RODENTS

Under wildly fluctuating conditions animals dependent on seeds, such as rodents, will usually have a hard time. Granivory in the Australian deserts seems to be a successful strategy only for highly vagile animals such as birds, or for social insects such as ants that can survive on reduced food supplies for extended periods and store seed in granaries. In this fundamental way, Australia's arid zone is unlike other deserts of the world in which granivory, particularly in rodents, is a successful way of life. There are, in fact, few desert-dwelling rodents in Australia and of those, most appear to be omnivorous, eating insects, roots and shoots as well as seed.

An ecosystem based on invertebrates, particularly termites, is expected to rammify into a web of invertebrate hunters, such as this predatory Carabid Beetle (below) and centipede (Alletheura sp.) ***(left). The iridescent Jewel Beetle is a miniature grazer (below left).***

COLLEMBOLA

Collembola, or spingtails (tiny, 0.5 to 1 mm long litter-hoppers) are often the most numerous arthropods, with as many as 200,000 animals per metre squared, in the leaf litter and top soil of Australian deserts. This is despite that fact that they need high humidities and their greatest diversity is in the humid climates. Those relatively few species that live in the deserts have developed special adaptations. One of the most curious of these is their ability to become anhydric, virtually dessicating like dried peas only to become plump and active again after a dousing in water. In fact Collembola can become active within one hour, after just a shower of rain.

The arid zone species of Collembola are litter detritivores under moist conditions and micro-grazers of algae and lichen in dry conditions. They have an important role in maintainance of the lichen/algae crust of the soil surface and in catalysing microbial action and hence the breakdown of litter. They do this by dispersing microbial propagules into other areas, maintaining microbial diversity through grazing and breaking the dormancy of microbial spores by their passage through the Collembolan's gut.

Scorpions

Scorpions, the oldest known terrestrial animal and maybe even the first to have conquered the land, are common in tropical and subtropical areas. But the custom, in the desert, of shaking out shoes is a wise precaution. In Australia, the arid zone correlates best with scorpion distributions and has more scorpion species than any other subregion.

This Urodacus *scorpion can grow to around 10 centimetres in length. A formidable hunter, it has 'conquered' the arid zone by evolving a deep, spiral burrow habit which maintains a humid environment.*

This is because the endemic genus, *Urodacus*, with 11 species, has abandoned the traditional scorpion sheltering places such as under stones or the bark of trees (and in shoes), and has evolved the deep, spiral burrowing habit which has enabled it to take advantage of the widespread arid conditions of Australia.

The uncommon scorpion, *Isometroides vescus,* goes about things a little differently. It occupies burrows built by others after first eating the owner. In fact this scorpion appears to prey exclusively on burrowing spiders.

Frogs

In frogs the adaptations to environmental extremes are simply the elaboration and perfection of an extremely common phenomenon. Thirty-five per cent of the Australian frog fauna burrow. Even in the red sand and spinifex plains of the Tanami Desert, after heavy summer rain, the land of scurrying lizards turns into a frenzy of leaping frogs -- up to 50 frogs per hectare.

Water-holding frog (Cyclorana platycephalus), *digging in for the day.*

This Trilling Frog (Neobatrachus centralis) *is another of the inland's water-holding frogs. Its name stems from the prolonged high-pitch trill the frog calls when advertising for a mate.*

A rare picture of the Tanami Desert in flood.

There are at least 20 species of burrowing frogs in the arid zone, and these so-called 'water-holding' frogs are quite distinctive. They usually have a broad head, bulbous body and short limbs, with structures called metatarsal tubercles, like little spades, on the undersurface of the feet which aid in digging.

Frogs do not have a waterproof skin. On the contrary, when they need water, they flatten themselves out like frog pancakes with belly down against any moist surface. The spaces between the cells of their ventral skin develop an increasingly negative pressure as water is lost and this pressure pulls water from the skin into the body. So, how do they insulate themselves during inevitable dry periods? Water-holding frogs develop an external cocoon derived from the normal frog process of sloughing off the old, dead outer coat (and then stuffing it into their mouths). In cocoon formation the outer layer of dead cells separates from the cells beneath but is not shed. Instead it sits around the mucous-coated frog like plastic film wrap, virtually impervious to water.

Frogs burrow to considerable depths and therefore a very heavy rainfall is required with enough volume for percolating water to reach an entombed frog. Up until very recently frogs were considered opportunistic breeders, ready to breed immediately any rain fell. However, it is more likely that frogs are seasonal breeders, responding to a lot of summer rain. At high temperatures the developmental rate, from egg to tadpole to frog, will be accelerated. The importance of this is immediately obvious watching torrential rain being swallowed up by sand deserts. Often puddles are nowhere to be found even though frogs, which apparently need puddles to breed, literally abound. Presumably they travel kilometres to the odd, very temporary pool. Or perhaps they have some other breeding tricks we haven't yet discovered.

***The ugly duckling of the water-holding frog world is this globular, warty-skinned and stubby-limbed* Notaden nichollsi.**

It takes a heavy summer rain and steamy night to turn on these water-holding frogs (Cyclorana maini).

There is one desert-dwelling frog, however, that does not need a puddle to breed in. The tiny, 3 centimetres long, Sandhill Frog *(Arenophryne rotunda)* lives, of all the very un-frog-like places, in the coastal sandhills of Western Australia near Shark Bay and, further south, at the mouth of the Murchison River. In winter males and females form pairs burrowing together and remaining below the ground for at least five months where mating takes place. As the sand dries the frogs move down following the boundary of the moist layer. The female lays 6–11 creamy white eggs in which the entire development occurs. The baby frogs emerge after 10 weeks.

Perhaps not surprisingly the diet of these frogs comprise the most abundant fauna of the sand dunes: ants. Anyone who has watched a frog hunt and eat will think that the only limitation on what they will eat is size; can it be stuffed into their mouth? Frogs will eat the resources available at the time, and after summer rain in spinifex grasslands there are abundant beetles, flies, spiders, grasshoppers and moths, but particularly termites and ants which form the major component of the frogs' diet.

THE FORWARD BURROWING FROG THAT BURROWS BACK-TO-FRONT

The Sandhill Frog *(Arenophryne rotunda)* lives in the coastal sand dunes of arid Western Australia. There it feeds on ants which are the most abundant insect life on the dunes. Unlike almost all other burrowing frogs in Australia that burrow with their feet, the Sand-hill Frog begins burrowing with its hands. Instead of using the normal backwards shuffling, this frog literally dives beneath the surface using its hands and feet as paddles. It apparently lies under the surface during the day and comes to the surface to feed at night. The significance of the burrowing is simply that about 10 centimetres below the surface the sand is moist and fresh.

NINGAUI

Ningaui is a genus of tiny carnivorous marsupials described only in 1975. Since then they have been recorded right across the arid zone where they live in spinifex grasslands. Despite their tiny size — they are considered adult when they weigh 5 grams — they are ferocious nocturnal insectivores often subduing prey bigger than themselves, such as grasshoppers, in a series of fast bites around the head.

The few studies that have been done on the animal show that it is not a social creature. It compensates with a loud and large vocal repertoire to communicate whether or not it wishes interaction. It breeds seasonally, during spring and summer and apparently does not live to breed again. Therefore the survival of the species depends on a single, annual generation of young.

Other little known species of carnivorous marupials are the Ooldea Dunnart *(Sminthopsis ooldea)* and the Long-tailed Dunnart *(Smininthopsis longicaudata)*. The Ooldea Dunnart is named after a small settlement on the Trans-Australian Railway where the first species were found in the 1970s. It has since been found regularly and will probably prove to be a common species. The Long-tailed Dunnart, however, is a rare animal found in rugged rocky areas sparsely covered by *Acacia* shrubs over spinifex. Prior to 1981 only 4 animals had been caught. In that year 9 animals were found in the Young Range, a series of flat-topped hills in the Gibson Desert Nature Reserve.

CHAPTER 4

CHENOPOD SHRUBLANDS

'You'll find gibbers and bloody sand,'
His mates had said.
ROLAND ROBINSON, 'The Wanderer'

Arid, wind-swept chenopod (Atriplex sp.) shrubland.

Salt-encrusted cushion plants near Dalhousie Springs, South Australia.

Saltbush and bluebush are the main components of the chenopod shrublands. Like the infertile sands which cover much of the arid zone, the chenopod shrublands are an extreme of another kind; an environment rich in nutrients, even to excess in some elements like calcium, magnesium and sodium, but with no surface water.

The sparseness of the chenopod shrublands, where separate plants can easily be counted, reflects their aridity but their succulence (a characteristic which is virtually absent from the rest of the sclerophyll-hardened arid zone) hints at their nutritional quality. Today, millions of kangaroos (and millions more sheep) confirm it.

The chenopod shrublands cover 434 000 square kilometres, or 8 per cent of the arid zone. They are composed of salt tolerant xeromorphic shrubs. At least 100 different genera of chenopods clothe a southern arc of inland plains and undulating lowlands where relief is generally less than 30 metres. Some saltbushes and bluebushes, such as northern bluebush (*Chenopodium*

The very stones lying upon the hills looked like the scorched and withered scoria of a volcanic region, and even the natives, judging from the specimen I had seen today, partook of the general misery and wretchedness of the place.

(Edward John Eyre, 1839.)

Gibber-hopper beautifully camouflaged against the copper-coloured gibbers.

auricomum), do grow in the northern areas, but these are restricted to ephemeral swamps and areas of aggregated drainage. Genera such as *Atriplex*, *Maireana*, *Sclerolaena*, *Chenopodium* and *Rhagodia*, each containing in excess of 30 species, characterise the chenopods.

Chenopods are found on a wide variety of soil types in all positions within the landscape, from skeletal soils of rocky uplands to saline drainage systems. Bladder Saltbush (*Atriplex vesicaria*) grows on clay soils, Pearl Bluebush (*Maireana sedifolia*), an extremely long-lived bush which has been found to be 300 years old, grows on medium-textured soils, and Black Bluebush (*Maireana pyramidata*) on calcareous sands and loams. Scattered trees, often Belar (*Casuarina cristata*) or Rosewood (*Heterodendrum oleaefolium*), and occasional woodlands are also a feature of these landscapes.

The density of the chenopod shrublands can be from sparse to over 500 perennial shrubs per hectare. In the normal extended dry period the ground is bare between the 1–3 metre high bushes, but after effective rainfall it is invisible under a carpet of ephemeral plants.

In fact Australia's economy, when it was riding on the sheep's back, produced the best of the golden fleece from these saltbush and bluebush plains. Banjo Paterson eulogised the plains with the unforgettable words: 'and he sees the vision splendid of the sun-lit plains extended . . .' These were the chenopod shrublands that were among the earlier-settled areas of the arid zone; the epitome of the outback.

If Eyre had paid attention to it, as he stood in 1840 at the northern edge of the Flinders Ranges in South Australia, surveying the unexplored landscape ahead of him, he would have noticed that herbivores did indeed abound with every step he took. Grasshoppers would have sprayed out with every footfall; they normally occur in the richer parts of the arid landscape and their presence in the saltbush and bluebush plains indicate that, contrary to their 'wretched' appearance, the chenopod shrublands are nutritionally rich. Eyre was also the second European to document the Stick-nest Rat, now extinct on mainland Australia.

Stick-nest Rat

There were two species of Stick-nest Rat on the mainland at the time, the Lesser Stick-nest Rat,* Leporillus apicalis, *and the Greater Stick-nest Rat,* Leporillus conditor. *These fluffy rats had compact bodies, long ears, blunt noses and a long tail, which in the case of the Lesser Stick-nest Rat had a white tip.

These saltbush and grass-covered rocky hills (below and right) along the Birdsville Track in north-east South Australia are habitat for the rare Chestnut-breasted Whiteface which nests amongst the bushes.

PENNY VAN OOSTERZEE

They were abundant across a southern arc coinciding with the chenopod shrublands. This habitat clearly provided a nutritionally rich, reliable food from which the rats could also obtain water.

That the rats were indeed herbivores is confirmed by the only remaining population of the Greater Stick-nest Rat (the Lesser Stick-nest Rat is almost certainly extinct) on two tiny islands off Ceduna in the Great Australian Bight. The two islands which make up the Franklin Islands are each about 2.5 kilometres long and 1.8 kilometres wide. Here about 1000 Greater Stick-nest Rats remain. The Franklin Island rats are entirely vegetarian, feeding on any succulent edible shrub and herb and their fruits. A major feature of the plants is their high water content; the ubiquitous Roly-Poly (*Salsola kali*), for instance, is 75 per cent water, and the two saltbushes *Rhagodia crassifolia* and *Enchylaena tomentosa* are 64 per cent and 60 per cent water respectively.

If not for the incredible housing that these rats built for themselves, Eyre and other explorers might not have noticed them. This typical nest was described in the Nullarbor in 1927:

. . . [the Stick-nest Rat] is remarkable for the fact that it builds for itself a wonderful home of sticks, and yet it has a burrow underground. The house is built of fine sticks, and these are so placed that they withstand the onslaughts of fierce winds. Frequently I have come across houses in which stones have been intermingled with the sticks. These stones, which are from one to one and half inches in diameter have been placed by the House-building Rats in among the sticks to give the structure a solidity sufficient to withstand the gales or the assault of enemies. Generally the homes are located under low tree branches, or in isolated hollows against some kind of shelter, and they are built of various dimensions, the usual size being about three feet in diameter and two feet in height. One large house that I examined measured six feet in diameter and was four feet high, compact, circular in shape, and tapered to the top . . . It was constructed of grass, leaves, stones and small sticks . . . In the interior, two feet above the ground, was a nest, in which there were young. The nest was made of grass, very fine sticks and leaves. Towards the exterior of the building, the sticks were comparatively large . . . Each year the original home was added to . . . and so in the course of years, the house is built larger and larger . . . They usually start work at dusk, and the greater part of construction is, therefore, done in the darkness . . . Immediately under the nest, in the centre of the ground floor of the house, there is an opening, which connects with their underground burrow . . . In the ground there are many holes leading into the nest, and connecting with the underground burrow.

(A.G. Bolam, 1927).

Red Kangaroo

The Red Kangaroo prefers the fresh green pick of grasses and herbs growing in between the chenopods after rain, and only switches to chenopods (such as Black Bluebush) during drought. It moves between sandplain and floodplain in response to vegetation or pasture growth, concentrating on the more palatable vegetation of the floodplain as the environment dries, but quickly moving onto the sandplains after rain.

Considered the most arid-adapted and hence nomadic of the kangaroos, in reality the Red Kangaroo has a small home range, usually less than eight square kilometres. Its wanderings seem to be restricted to this movement between soil type although for reasons unknown a few individuals occasionally range more widely.

During the frequent droughts experienced in the chenopod shrublands, which are the driest regions of the country, a sedentary existence probably has some advantages. It makes more sense to know where shelter and food are rather than to spend time searching for greener pastures, particularly if stressed. As extensive areas become denuded, kangaroos move within their home range to gather on remnant pasture. The dramatic increase in kangaroo density creates a false impression of high population levels. As the remnant pasture dwindles, the kangaroos die. The sudden decrease of kangaroo density is often mistaken for mass exodus. What it really means is mass mortality.

The Red Kangaroo functions under a very simple system; the more kangaroos, the more pasture is eaten, the less pasture is available, the fewer kangaroos remain. The rapidity with which effect follows cause in this system is quite staggering: the pasture responds to rainfall within a few days; the Red Kangaroos conceive within two weeks; they disperse to different soil types where their condition improves within a month; the number of kangaroos increases before six months are out.

The driving variable, to which Red Kangaroo populations are closely tied, is rainfall. Kangaroo populations go up and down according to the effect of rainfall on plant growth. The chaos which seems to characterise Red Kangaroo numbers in the outback is really a reflection of the fractal patterns of rainfall.

One of the largest living marsupials, the Red Kangaroo (Macropus rufus) prefers the open plains. It requires some shade, which it obtains from scattered trees. It is mainly nocturnal and can be found sitting in the shade of trees during hot days, or sunbaking on mild days. If enough green herbage is about, the Red Kangaroo does not need to drink fresh surface water. Not all Red Kangaroos are red. Most males are reddish, but some blue-grey males occur. Females are mostly blue-grey in colour, but can sometimes be reddish also.

GARY STEER/SKY VISUALS

Saltbush

In terms of distribution, the saltbush genera* Atriplex *is one of the most important genera in the arid zone because it forms dominant perennial shrublands and has annual and perennial members which grow in a wide range of communities.

A saltbush (Atriplex *sp.*) *plain. The genus* Atriplex *is not endemic to Australia, although 70 per cent of its species are. It was possibly distributed across the continents during Gondwanan times.*

R.D. GRAETZ

Despite this, *Atriplex* is difficult to identify. The common Bladder Saltbush, *Atriplex vesicaria*, for instance, has many geographic and edaphic (related to soil type) variants. It seems to be a very 'plastic' species, even changing shape quite considerably under different levels of grazing.

The genus *Atriplex* is not endemic (however, 70 per cent of its species are). It occurs worldwide, possibly being distributed across the continents during Gondwanan times. The thinking of some scientists is that it arrived in the Geraldton–Carnarvon area of Western Australia in a well-defined state, and evolution of the genus continued from there in isolation from the genus in other places.

The presumed parental *Atriplex* still exists in the coastal Shark Bay region of Western Australia. It is an entirely bladderless form of the Bladder Saltbush (so named because of the inflated, hollow bladder which encloses the seed). From this bladderless parent, the Bladder Saltbush evolved through the small bladder form — the distribution of which stretches across the Nullarbor Plain — to small and large bladder forms which inhabit a variety of heavier soils throughout southern arid regions, including stony soils and saline clays.

Old Man Saltbush (*Atriplex nummularia*) is widespread on heavy clays in the arid zone and includes a subspecies, *spathulata,* which inhabits arid loams of the Nullarbor. Centres of diversity for *Atriplex* today include its original landfall in Australia (the area around Shark Bay), noted for its high species diversity, and in the ranges which interrupt the broad plains of South Australia. These are regions where winter rainfall is the most reliable even though heavy rain does sometimes fall in summer. As with the other major plant groups of the arid zone, the unpredictability of the climate allows a correspondingly diverse array of life-history strategies to be used by different chenopods.

The Bladder Saltbush (*Atriplex vesicaria*), for instance, grows on the more clayey, salty soils which do not allow water to easily filter through. Its roots, accordingly, are shallow, and spread out like a fan for as much as two metres around the plant, so that they are able to collect as much water as possible from the surface layer of soil. This water is relatively fresh because it has not penetrated any depth of saline soil. The plant grows rapidly when moisture is available. When the soil is dry, the Bladder Saltbush shuts down and has the remarkable ability to control the aperture of its stomates. In this way evapotranspiration can be reduced by 50 per cent. In dire straits the plant will even drop its leaves and shed fine roots to save energy (a rare strategy for Australia's desert plants where growth on normally sterile soils is hard won).

The Bladder Saltbush flowers and sets seed after rain, no matter when it falls. The seeds, however, germinate best when the temperature is low and when rainfall is spread out over days, which is precisely what winter rainfall is likely to do. The seeds have an armoury of germination inhibitors, making them viable for some years. The envelope which surrounds the fruit is impregnated with an inhibiting salt. It takes 50 millimetres of rain over several days to leach this salt. On top of this, a hard seed coat and a light inhibitor, thrown in for good measure, also delay seed germination. The plant itself lives for about 25 years. This is relatively short-lived, so the plant needs to make sure that its seeds have the chance to grow into young plants. The seeds have to be able to wait around for the right conditions.

In contrast to the Bladder Saltbush, the Black Bluebush (*Maireana pyramidata*) grows on sandier soils and, therefore, has extremely deep roots. Sandy soils, like those of the spinifex grasslands, allow water to percolate more easily. Because the Black Bluebush is able to tap the deep — and hence more stable — water supplies, it tolerates a higher and greater temperature range than does the Bladder Saltbush, and it grows at a steadier rate. In contrast to the Bladder Saltbush, it only loses its leaves when it is dying.

The Black Bluebush flowers mainly after autumn or winter rains. It sets seed in summer and the seeds, which only survive for a few months, need only one or two heavy rainfalls to germinate; precisely the sort of rain likely to occur in summer. The seeds of the Black Bluebush do not have the inhibition armoury of the Bladder Saltbush and so lose viability in a matter of months. The Black Bluebush plant itself has a life span of 150 years or more; a long time when compared to the Bladder Saltbush.

The plant gambles on the fact that at least one or two rare summer deluges will occur in its 150 year life span. In contrast to the Bladder Saltbush, the Black Bluebush invests its energy in the adult plant rather than the seeds.

Compared to the other ecosystems of the arid zone, chenopod grasslands are ecologically simple because of the absence of trees and the minor involvement of fire. All major research into the shrublands has focused on the dynamics of the shrubs themselves, particularly in relation to grazing by sheep. Astoundingly, apart from reports involving single visits, there has been no long-term study of the wildlife of the chenopods. Nevertheless, from what reports do exist we can see that this ecologically simple system supports significant populations of a characteristic wildlife.

This Pop Saltbush **(Atriplex holocarpa)** ***has inflated, hollow bladders which hold the seed.***

Breakaway Country

If Eyre thought of the stony slopes north of the Flinders Ranges as 'scorched and withered scoria', heaven knows what he would have thought about the breakaway country further north. Formed in the lowest parts of Australia, it is a baffling and surreal landscape where active creeks are biting into the ancient Tertiary plains, making a flat country even flatter.

The Fat-tailed Dunnart* (Sminthopsis crassicaudata) *uses the energy reserves in its tail when food becomes short.

It is a place of ochre-coloured mesas crumbling into extensive gibber footslopes and plains. A place where resistant ridges and escarpments spill run-off onto ephemeral outwash fans and floodplains.

On the chenopod covered footslopes of the breakaway country, the limestones and mudstones of the Tertiary have weathered to a deep cracking clay. These cracks act as ready-made burrows for animals such as the Narrow-nosed Planigale (*Planigale tenuirostris*), Fat-tailed Dunnart (*Sminthopsis crassicaudata*) and Forrest's Mouse (*Leggadina forresti*). The Narrow-nosed Planigale and the Fat-tailed Dunnart are carnivorous marsupials attacking insects their own size such as moths, beetles and grasshoppers which provide moisture as well as food. Like a trophy-hunter, the Narrow-nosed Planigale lines its nest with the insect wings of its prey.

There is an entirely different suite of invertebrates in the chenopods than in the cellulose-rich spinifex grasslands and mulga woodlands. Termites, for instance, are few and usually found only in the creeklines where there is woody vegetation. Herbivorous grasshoppers, moths and beetles are abundant, as are spiders preying on them and, of course, the ubiquitous ants.

The ants not only interact with other animals but they also interact with the plants. Many chenopods are myrmecochorous: relying on ants for seed dispersal. The plants provide a food body to which ants are attracted. But the food body is not easy to extract, so the ant drags the whole fruiting structure containing seeds, fruit and woody casing, back to its nest. Here it extracts its reward and discards the woody casing with its undamaged seed contents into refuse piles covering its nest mound.

This of course is just what the plant 'wants'; the ant mounds are favourable micro-sites for germination and growth. Because of the 'rubbish' discarded by ants — scraps of invertebrate carcases and other detritus — the mounds are relatively nutrient rich. Also the soil around mounds is often less compacted so that water more easily penetrates. Plants such as some of the copperburrs (*Sclerolaena*) grow in association with ant mounds. And some specific plants such as the Green Copperburr (*Sclerolaena diacantha*), with large fruiting bodies which smaller ants find difficult to haul, grow exclusively on the mounds of the large ant *Rhytidoponera*.

Baldwin Spencer took an interest in the fruiting structures of the copperburrs while on the Horn Scientific Expedition to central Australia in 1894:

The seed-cases of these have a pretty downy centre, perhaps half an inch in diameter, but around this are a number of very stiff, sharp-pointed spikes projecting through the soft down. What with these and other prickly seeds our camping place was often a bed of thorns, and after selecting a spot, a usual preliminary to opening out our rugs was to sweep the ground with an impromptu broom of Cassia *branches.*

The reason the casing around the copperburrs fruit is modified into spines is a bit of a mystery. Spencer attributed it to a climatic response, rather than to a protection against herbivores. But he added that if animals want to feed on these 'climate-proof' plants, then they must become fitted to do so.

Silcrete forms protective crusts on the flat-topped hills or mesas that characterise dry deserts worldwide. These crusts gradually break away around the edges, giving rise to the colloquial term for this country: breakaways.

MEGA-FAUNA

Near where Spencer mused on what might have eaten the unforgiving thorns lay fossilised bones of some of the giant, now extinct, mega-fauna. Possibly two species of *Diprotodon*, a giant wombat, and at least five large species of kangaroos became bogged, perhaps only 20 000 years ago. And these would have been only a small sample of the incredibly diverse fauna of the late Pleistocene; a fauna which clearly must have been 'fitted' to feed on the thorny, but nutritious copperburrs.

Large kangaroo skeleton — the last of the mega-fauna.

ECHIDNAS

Echidnas live in the limestone caves of the breakaway country north of Coober Pedy. Also sometimes found in these caves are remnant nests of a once abundant Stick-nest Rat population. In fact most nests that remain on the continent today are found in the protective breakaway systems.

Dragon lizards are endemic to Australia. The Nullarbor Dragon (*Pogona nullarbor*) is a moderately large 'bearded dragon' restricted to the Nullarbor Plain.

LIZARDS

Lizards occur in the chenopod shrublands but, predictably, nowhere near as abundant as in the spinifex grasslands where they flourish within a foodweb based on termite tidbits. Particularly common are geckos and the skink Broad-banded Sandswimmer (*Eremiascincus richardsonii*) which is found where fissures and gilgais (depressions or basins forming a hummocky micro-relief found in clayey soils) are found in the soil. Dragons such as the Central Bearded Dragon (*Pogona vitticeps*), the Nullarbor Bearded Dragon (*Pogona nullarbor*), and the Earless Dragon (*Tympanocryptis* spp.), as well as the widespread goannas, the Perentie (*Varanus giganteus*) and the Sand Goanna (*V. gouldii*), are also found in breakaway country. Many of these animals are focused on the more reliable wooded footslopes of mesas and the woodlands of the ephemeral streams. Birds in particular are focused on the Coolibah (*Eucalyptus microtheca*), Red Mulga (*Acacia cyperophylla*) and Gidgee (*A. cambagei*).

Right: The Central Bearded Dragon (*Pogona vitticeps*) is widespread, preferring a semi-arboreal lifestyle. All dragons are egg-layers —this one lays up to 30 eggs.

Left: The Short-beaked Echidna (*Tachyglossus aculeatus*) is found right across Australia. Except for a wholesome diet of ants and termites, it has no particular habitat preference. When disturbed, the echidna rolls itself into a prickly ball.

Gibber Deserts

Even the sun and sand-blasted gibber (the word comes from an Aboriginal word meaning 'stone') deserts have their own characteristic suite of animals.

Most characteristic of gibber deserts is Sturt's Stony Desert, perhaps the most inhospitable place on the surface of the continent. Walking through it in summer is every bit as hellish as Sturt found it in 1845 when the 'heated and parching blasts' that came upon him were more than he could bear.

Yet several animals are largely restricted to stony terrains. These include the Earless Dragon, the Kultarr (*Antechinomys laniger*), and the larger predatory Kowari (*Dasyuroides byrnei*), a carnivorous marsupial with a striking black-tufted tail which will devour other mammals such as the Long-haired Rat (*Rattus villosissimus*). Some bird species also favour gibber deserts; the Chestnut-breasted Whiteface, Thick-billed Grasswren and Western Fieldwren are often found in areas with a moderate cover of taller bluebush and saltbush. The Gibberbird prefers the barer stony plains. And so, extraordinarily, do the Inland Dotterel and, seasonally, the Australian Pratincole; extraordinary because these birds, which are adapted to the driest and most irradiated part of Australia, are in fact seashore or aquatic species.

Unlike the Australian Pratincole, which is nomadic and migratory, and the dotterel which is also nomadic, the Gibberbird is virtually sedentary. This specialised ground-living honeyeater is therefore considered Australia's most complete desert bird. Yet not much is known about it.

All three birds nest on the ground. The Gibberbird builds a cup of dried grass and other material, the Inland Dotterel's nest is a scrape lined with any material including stones, sticks and sand, while the Australian Pratincole just lays its eggs on the bare soil amongst the gibbers.

L. ACRAMAN

A shallow scrape in the bare ground is home for these neatly camouflaged, self-shadowed Dotterel eggs.

Sturt's Stony Desert, near Innamincka.

The Australian Pratincole and Gibberbird nest in spring and summer. The former never leaves its eggs while the sun is out and sits very close on the eggs. Clearly, the bird's body is much cooler than the ambient temperature. (Under these conditions, a thumb placed on the bulb of a thermometer would make the mercury plummet.) Only in overcast conditions or when the ground is in shadow do the birds leave the eggs — when the sun reappears they quickly return. The Australian Pratincole's clay-coloured chicks are so well camouflaged that they become almost invisible on the stony plains.

The Inland Dotterel is also a seasonal breeder but nests in winter because it is at least partly herbivorous and its habitat is the chenopod shrublands where winter rainfall is more reliable for plant growth. The Australian Pratincole and Gibberbird are insectivorous and their breeding coincides with the time of more abundant insects.

Country like these bluebush and saltbush plains was the home of the now-extinct (on the mainland) Stick-nest Rats.

Hamersley Ranges, Western Australia.

Amazingly the Australian Pratincole rarely seeks shelter, even when the temperature in the shade is 46 degrees Celsius. It sometimes stands with its bare legs in the shade of a bush but this is probably incidental to the bird avoiding the unbearably hot stones. Not that this seems to be a problem because on hot days the bird walks unbelievably slowly over the gibber with a high-stepping gait. Perhaps in this way one foot cools in the shadow of the bird while the other foot parboils on the 70 degree Celsius stones. To compensate for this incredible ability to tolerate heat the Pratincole needs to pant constantly and drink often. It can tolerate the brackish fluid of the salt pans by excreting excess salt through the salt glands above its nostrils. The chicks of the Australian Pratincole do not drink because their nest is usually built some two kilometres from water. They appear to overcome the need for water by resting throughout the day deep within bushes (which are about six degrees cooler than in the shade of hides made by scientists studying them!).

The Inland Dotterel, sensibly, seeks shade at around 46 degrees Celsius. It loses a lot of water through evaporative cooling while panting but does not actually drink, obtaining all its water requirements from eating plants during the day. It also has a salt gland through which salt is excreted. The Inland Dotterel is interesting in that, while it is a diurnal herbivore, it switches to an insectivorous diet at night. Probably the water-supplying plants do not provide enough energy for living.

The Inland Dotterel must be one of the world's most perfectly camouflaged ground birds. In the open the birds turn their backs on intruders and crouch motionless so that their flecked plumage merges with the background. If pursued behind bushes, they turn face-on and the black lines on their chest miraculously blend with the plant stems.

Other birds, such as the Bushlark, occur in gibber deserts seasonally and are attracted to the richer vegetation found in gilgais. In the Nullarbor similar depressions are called 'dongas'.

HOW GIBBER FORMS

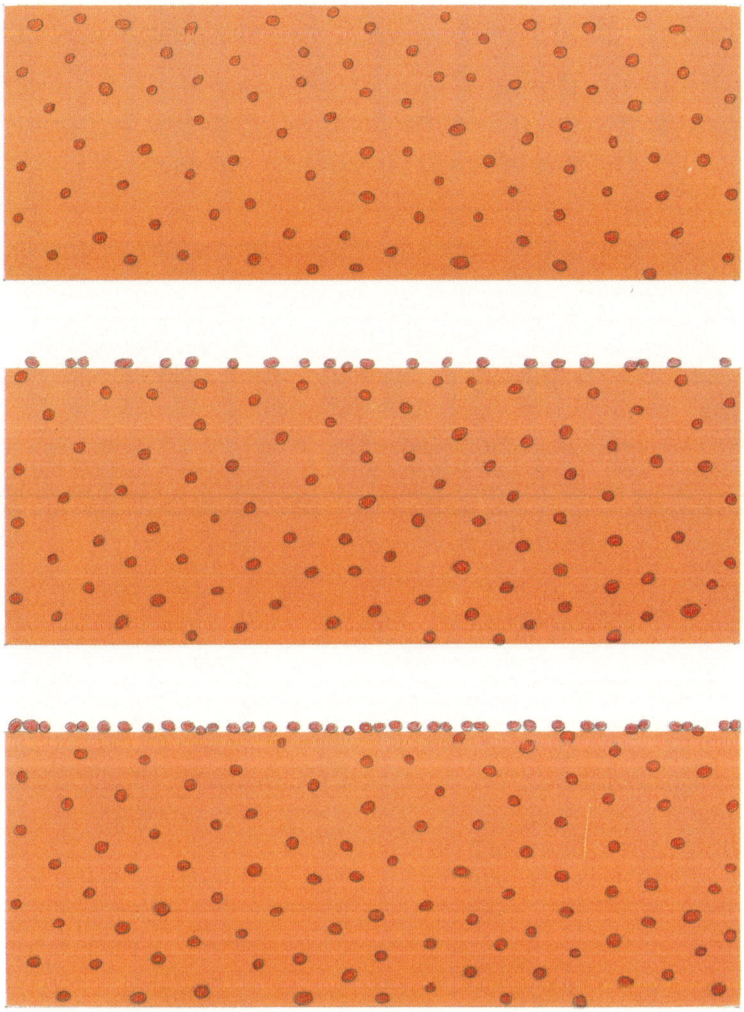

'Throughout all this district the low flat-topped desert sandstone hills indicate the original level of the land. All these hills have a thin capping of hard chalcedonised sandstone; when once this is broken up the softer underlying rock is rapidly disintegrated, and the sand particles into which it breaks up are partly carried away in flood time, and partly blown away by heavy winds. The harder chalcedonised material gradually breaks up into blocks of various sizes, and these become polished and rounded by the wind-blown sand grains, while a thin coating of oxide of iron gives them a red-brown and curiously polished appearance. As the sand is gradually removed the polished stones come to form a layer spread over the flat surface of the plains, the stones of which are so close to one another and so regularly arranged that at times they look almost like a tesselated pavement. In passing from the plains up the sides of the hills the gibbers can be seen in all stages of formation, from the small, smooth and flattened pebble on the plain to the big, irregularly shaped mass which has just tumbled off from the exposed surface of the thin desert sandstone capping of the hill.' (Baldwin Spencer, 1896).

The Nullarbor

Dongas (from the Zulu word* adonga*) are characteristic of the Nullarbor limestone often leading to sink holes or caves. They are typically circular, a few metres to a few hundred metres in diameter, and a little more than a metre deep. They support a distinctive vegetation because of the local concentration of run-off.

Twenty million years ago, this would have been a sea floor. Limestone deposited by marine organisms has weathered to form cemented sheets and nodules of calcium carbonate.

Interestingly, of the two birds restricted to the Nullarbor, the Nullarbor Quail-thrush and the Naretha Blue-bonnet, the latter, on the treeless plain, is associated with the dongas.

At 250 000 square kilometres, the Nullarbor is the largest karst (limestone characterised by caves and underground drainage) area in the world; a 10 million year old sea-floor pushed high and dry. The word Nullarbor is derived from the Latin *Nullusarbor* meaning no trees. There are no trees because the soil, derived mainly from sea-shells, is a weakly developed, shallow calcium-rich loam. The treeless plain is an oddity in an arid zone characterised by its infertile soils and patchiness. It is in fact much more like the deserts of other parts of the world, with its succulent and semi-succulent vegetation. It is incredibly uniform and covered by widespread saltbush and bluebush shrub species of the genera *Atriplex*, *Maireana*, *Sclerolaena*, *Lysiana* and *Rhagodia*. There is a more diverse annual component, such as grasses and herbs, in certain areas such as the dongas, where moisture concentrates.

Predictably lizards are very poorly represented in an environment characterised by relative nutrient richness and horizontal uniformity. The average species diversity of lizards in a region of the Nullarbor is around 4, compared to 39 lizards in the Great Victoria Desert to its north.

Again, like deserts in other parts of the world, such as in North America, the Nullarbor had, until recently, a surprisingly abundant mammal fauna. Prior to European occupation, 32 native mammals were known to inhabit the Nullarbor, with the rodents in particular being remarkably abundant — in 1922, up to six Stick-nest Rat nests were counted per donga. Today most of these animals have given way before an explosion of rabbits, cats, foxes and house mice.

Around 10 million years ago, the Nullarbor was raised high and dry above sea. Its flatness attests to the stability and low rainfall of the region since then.

CHAPTER 5

DESERT RIVERS AND SALT LAKES

Poor Charles Sturt. He was marooned at a rocky water hole on the edge of the parched and blistering stony desert later named after him. His men were dying of scurvy and his boat lay splitting and fading in the desert heat; even the visionary inland sea in his head was evaporating.

We had witnessed migration after migration of the feathered tribes, to that point to which we were so anxious to push our way. Flights of cockatoos, of parrots, of pigeons, and of bitterns; birds, also, whose notes had cheered us in the wilderness, all had taken the same high road to a better and more hospitable region.
(CHARLES STURT, 1845)

He was imprisoned for six months in temperatures that reached 61 degrees Celsius. Rain finally released him and he wearily pushed north-west, still searching for an inland sea near the centre of the continent.

After several forays west, north-west and a final push north, he came upon a magnificent channel with a large sheet of water dotted with waterbirds. So, this was the more hospitable region to which the birds had flown. And the region was populated with well-fed Aboriginal people too. Nevertheless, Sturt was at a loss to account for the numerous watercourses and astonished by the fact that one of his colleagues pulled a dozen gleaming fish from a cool, brackish pond. The country was, after all, thousands of kilometres from high land where he would have expected to find streams.

Before Sturt and his party returned to civilisation, he named the fine watercourse Cooper Creek after Judge Charles Cooper of South Australia. He added that 'I would gladly have laid this creek down as a river, but as it had no current I did not feel myself justified in so doing'. And in so saying Sturt unwittingly crossed and named the continent's second largest inland river system (after the Murray), terminating at Lake Eyre.

The inland sea: Lake Eyre south, full of water.

Above, dust storm south of Coober Pedy.

Where the creeks run dry or ten foot high
And it's either drought or plenty.
A.B. (BANJO) PATERSON, 'The Overlander'

Below, times of plenty: Lake Apanburra, Simpson Desert.

L. ACRAMAN

FOOD WEB OF LAKE EYRE

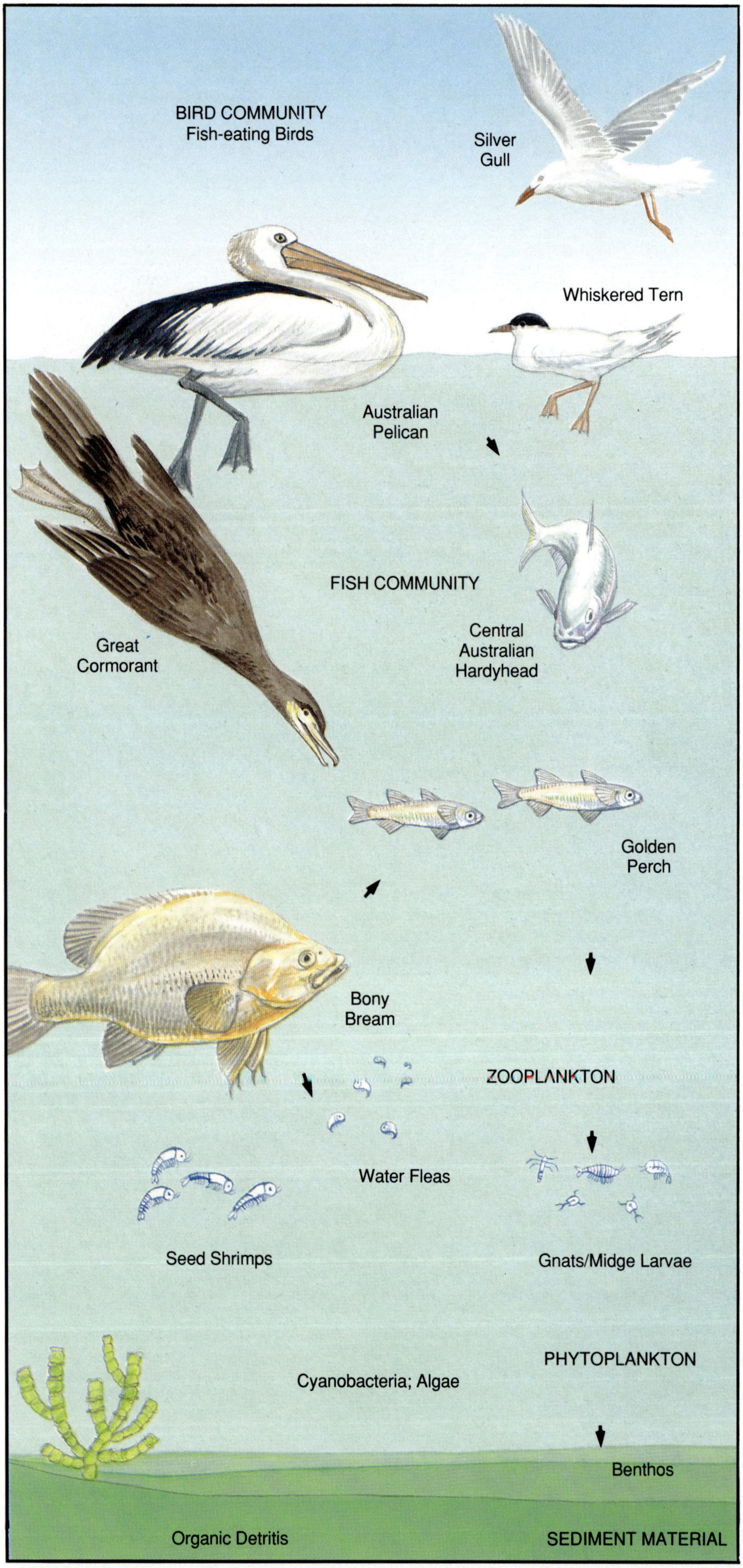

'. . . one, vast, low and dreary waste . . .' (E. J. Eyre, 1840)

Lake Eyre is one of the great natural wonders of the world and much has been written about the spectacle of flood when the so called 'dead heart' comes to life. Floods filling the Lake three times this century, during 1949–52, 1974–78, and 1984–85, have received a lot of attention. Much less scientific attention, however, has been given to the aquatic biota. A little information was gained in the 1949–52 floods and a bit more during the next two floods.

The information reveals a simple foodweb based on species which have an ability to scatter widely and quickly colonise favourable habitats. The few species of algae and bacteria which occur in Lake Eyre, for instance, also occur in other salt lakes. The same is true for the zooplankton. Of the larval stages of insects in the waters of Lake Eyre, larvae of midges, brine flies and other flies are common. The composition of insects, however, is different to other salt lakes in southern Australia. In Lake Eyre only those insects that can actively disperse (fly) and which can survive the great periods of desiccation (usually with highly resistant eggs which blow around in the desert dust) characteristic of the Lake are common.

Of the 'freshwater' fish able to survive in Lake Eyre, only the most hardy, the Bony Bream, Lake Eyre Hardyhead and Yellowbelly do well. All have an ability to tolerate broad temperature ranges and water as salty or saltier than sea water. The Bony Bream and Lake Eyre Hardyhead have a catholic diet and the Yellowbelly is a carnivore, no doubt eating the other two.

The top of the food chain is occupied by fish-eating birds, chiefly pelicans, silver gulls, great cormorants and whiskered terns. These birds were also the commonest species recorded on Lake Eyre during the last two floods.

Lake Eyre has a simple foodweb based on widespread species which can quickly colonise favourable habitat.

When full to brimming, Lake Eyre is fresh to brackish. Millions of fish flood into it. As the water evaporates, salinity levels increase. Fish species die out one at a time, according to their tolerance of salt. Eventually the water dries and salt blankets everything.

A landscape of salt and mirage.

Channel Country

In terms of rainfall the Lake Eyre region is the driest place in Australia, averaging about 100 millimetres of rain per year, barely enough to support even a meagre plant growth.

Right: Landscape so flat that the rivers lose their definition and fan out into multiple, intertwined channels over as much as 150,000 square kilometres.

Only the deepest holes are permanent. Most water is sucked into the parched air. Mud cracks in a dry claypan.

The lower the rainfall in Australia, and generally in 'tropical' deserts around the world, the more variable in both space and time it will be. For instance, in January 1974, in the vicinity of Cooper Creek, 696 millimetres fell at Innamincka, 319 millimetres at Cordillo Downs about 120 kilometres to its north, and 439 millimetres at Moomba, a short distance to its south. In January 1917, 23 millimetres fell at Cordillo Downs and 40 millimetres at Innamincka.

Yet this region is the largest refuge in Australia's deserts because run-off occurs across an extraordinary 1 300 000 square kilometres from the Great Dividing Range through vast, flat plains and into the sump of Australia; Lake Eyre itself at 15 metres below sea level.

The Diamantina and Cooper, the major tributaries of Lake Eyre, traverse country so flat that the rivers lose their definition and fan out into multiple intertwined channels over as much as 150 000 square kilometres. This vast natural irrigation scheme makes the Channel Country one of the richest parts of Australia. It contains islands of luxuriance within the rippling sand dunes of the Simpson Desert, and stretches of blazing stone in Sturts Stony Desert to its north and Strzelecki Desert to the east.

Regular flood pulses come down through the Channel Country from the seasonal tropical downpours in north and central Queensland. Despite these and the episodic major floods, only the deepest holes in the Cooper are permanent. Most water is sucked into the parched air. The evaporation rate here is one of the highest in the country, about 3600 millimetres (3.6 metres) a year.

Nevertheless the Channel Country has an incredible spectrum of wetland habitats. This has a crucial bearing on the colourful array of organisms and the strategies that they adopt to cope with impermanence.

Most flows along Cooper Creek do not result in inundation of the pan and lake-sprinkled floodplain. The most distant pans are usually dry. These may take the form of salt pans or bare claypans, but mostly they are ephemerally vegetated lake beds with heavy deeply cracking clay soils. Coolabah-lined backwaters, on the other hand, are shorter channels attached to the main channel of the Cooper. They regularly receive water as the Cooper rises and they also drain water from the flood-plain back into the Cooper as the flood wave passes.

The most biologically significant floodplain features, however, are the more permanent water bodies; the main channels of the Cooper Creek, and the Coongie Lakes complex which is at the terminus of the north-west channel of the Cooper. The most sizeable stands of timber in the region, including biologically important stands of river red gum, fringe these channels and lakes. Equally significant is the Tirrawarra Swamp on the north-west channel of the Cooper (which feeds its water into Coongie Lakes), the Embarka Swamp on the main channel of the Cooper and Goyder's Lagoon on the Diamantina. These swamps are large, densely vegetated, and contain strings of deep waterholes.

In reality it is these three immense swamps — Tirrawarra Swamp, Embarka Swamp and Goyder's Lagoon — that are the regular termini for Channel Country floods. Floods, in fact, rarely reach Lake Eyre — perhaps one in thirty, or two or three times per century.

Cooper Creek, for instance, branches 30 kilometres south of Innamincka and carries most floods along the north-westerly branch to Tirrawarra Swamp and then to the Coongie Lakes complex. Only in major flooding events do waters extend any distance down the main branch to Lake Eyre. And then the expansive Embarka Swamp mostly swallows this, being a major block point to flow along this branch.

Once in the swamp the main channel disintegrates. Rarely does this reappear downstream of the swamp to spill into Lake Eyre. When this does happen even gibber plains, sandy rises and dunes are flooded as water pours out into the interdune corridors and flats.

COOPER CREEK FROM SPACE

Coongie Lakes Region

Of the internationally significant wetlands of the Channel Country, Coongie Lakes, because of their relative permanence, are the most important. Enigmatically situated in the heart of a searing salty desert, they are four times fresher than the middle reaches of the Murray River.

The Coongie Lakes complex is the usual terminus of Cooper Creek floodwaters. Lake Goyder is the final lake of the lake complex to flood.

The yearly flood pulses from rain in the upper catchment flow into Coongie Lakes via the north-west branch of Cooper Creek and Tirrawarra Swamp. The water flows into the system of linked chains, first filling the three southern lakes, Coongie, Marroocoolcannie and Marroocutchanie (total 25 square kilometres), and then the two northern lakes, Toontoowaranie (13.3 square kilometres), via Brown Creek, and finally Goyder (39 square kilometres) via Ella Creek.

North of Lake Goyder is a high sand dune complex which stops water from flowing into the string of lakes to the north and north-east. These outlying lakes only fill with run-off from Sturts Stony Desert to the north-east after intense local rain. They therefore have a different aquatic fauna associated with them. For a start, because they are not connected to the Cooper system, they do not have any fish.

Pelicans descend on the lakes in spirals of a thousand birds.

Coongie Lakes are one of Australia's most important wetlands. Up to 35 000 waterfowl occupy the lakes all year. Mammals, frogs and one species of a tortoise are also abundant.

Prior to European settlement, the region was inhabited by hundreds of Aboriginal people, who relied on the exceptional resources of this almost miraculous oasis.

The north-west branch of the Cooper, with its associated Tirrawarra Swamp and the Coongie Lakes complex, has a great diversity of aquatic habitats. Yet the fish fauna is depauperate; particularly considering a catchment of 300 000 square kilometres. The reason for this undoubtedly has a great deal to do with the unpredictability of the environment; even the Coongie Lakes, at two metres deep, will dry out within 7–9 months if the Cooper stops flowing.

Out of the 26 or so species of native fish in the Lake Eyre Basin, 13 species occur in Cooper Creek. Chemically and physically, Australian inland waters are far from stable. Temperatures can range from 3 degrees Celsius on a winter night to over 40 degrees in the middle of a summer day. Keeping in mind that sea water has a salinity of about 35 parts per thousand, salinities of inland waters can range from fresh at 1 part per thousand, to hyper-saline at 350 parts per thousand. Dissolved oxygen, similarly, ranges from stagnant at 0 per cent to super-saturation at 192 per cent. This range encompasses the different natural waters and can also be found in a single water body over time; salinity can vary over the range in two to six months and oxygen from 15 per cent to 80 per cent, daily.

The Cooper Creek Floodplain

The floodplain of Cooper Creek has an incredible spectrum of wetland habitats. The usually dry peripheral salt pans and clay pans have heavy clay soils which are either networked with cracks and slabs of curling mud or covered in ephemeral herbs. The Coolibah-lined billabongs on the other hand are secondary channels which regularly receive water as the main channel of the Cooper rises. Billabongs also drain water from the floodplain

Claypan

Shield Shrimp
(Triops australiensis).

Dunnart
(Sminthopsis youngsoni).

Neobatrachus centralis

Royal Skink
(Ctenotus regius)

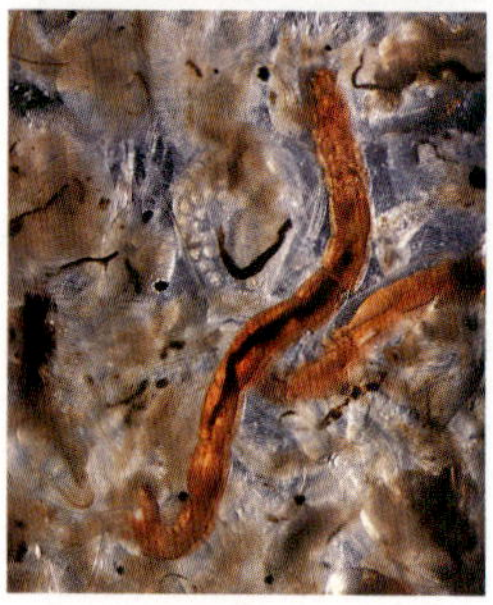

Chironomid larvae
(red), Fairy Shrimp etc,
L. Apanburra (zooplankton).

Flooded claypan

Snake *(Pseudechis)*

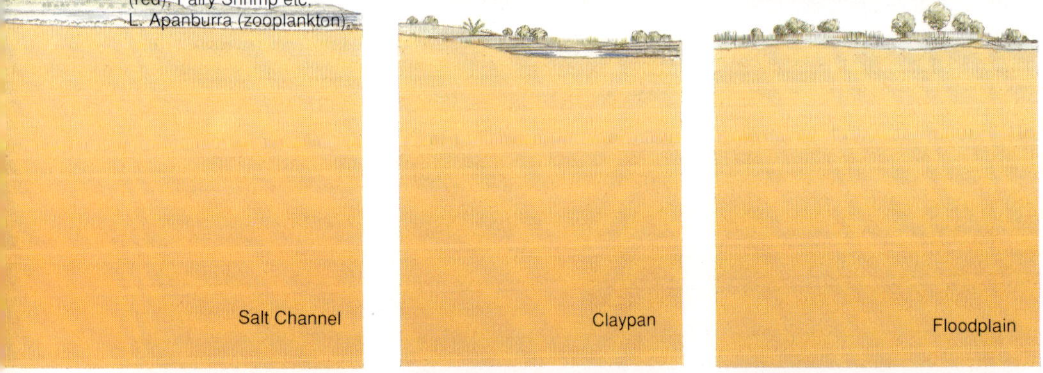

into the Cooper. Permanent water bodies such as pools in the main channel or the Coongie Lakes complex are the most biologically significant with their fringing stands of River Red Gums. But even these have a fluctuating water level. Coongie Lakes, for instance, will dry out within 9 months if Cooper Creek stops flowing. Sand dunes flank Coongie Lakes.

Brolga
(Grus rubicundus)

Wedge-tailed Eagle

Krefft's Turtle
(Emydura krefftii)

Palaemonidae
(Macrobrachium sp.*)*

Spinifex Hopping Mouse
(Notomys alexis)

Seed germination

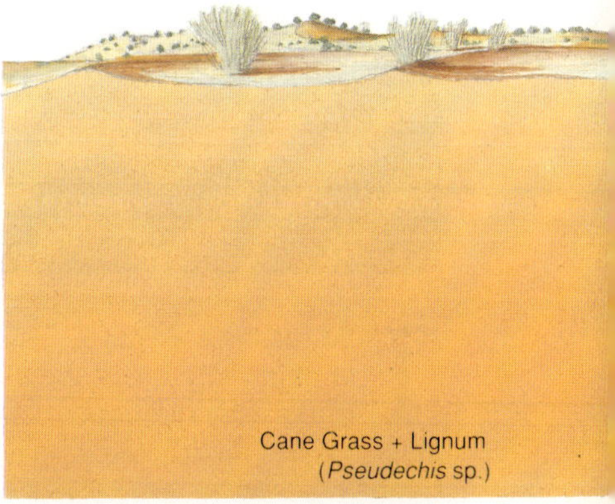

Fish of the Centre

The north-west branch of the Cooper, with its associated Tirrawarra Swamp and the Coongie Lakes complex, has a great diversity of aquatic habitats. Yet the fish fauna is depauperate; particularly considering a catchment of 300 000 square kilometres.

Only the most adaptable fish, such as this Central Australian Catfish, can survive inland waters where salinities vary from fresh to 350 parts per thousand (sea water is 35 parts per thousand) and where oxygen fluctuates from 15 per cent to 80 per cent daily.

The reason for this undoubtedly has a great deal to do with the unpredictability of the environment; even the Coongie Lakes, at two metres deep, will dry out within 7–9 months if the Cooper stops flowing.

Out of the 26 or so species of native fish in the Lake Eyre Basin, 13 species occur in Cooper Creek. Chemically and physically, Australian inland waters are far from stable. Temperatures can range from 3 degrees Celsius on a winter night to over 40 degrees in the middle of a summer day. Keeping in mind that sea water has a salinity of about 35 parts per thousand, salinities of inland waters can range from fresh at 1 part per thousand, to hyper-saline at 350 parts per thousand. Dissolved oxygen, similarly, ranges from stagnant at 0 per cent to super-saturation at 192 per cent. This range encompasses the different natural waters and can also be found in a single water body over time; salinity can vary over the range in two to six months and oxygen from 15 per cent to 80 per cent, daily.

Only the most adaptable fish can survive these fluctuating conditions. Families such as the eel-tailed catfishes, hardyheads and perches (or grunters) can do it. Arid Australia's most successful fishes are mostly from these families and include the widely dispersed and abundant Bony Bream (*Nematolosa erebi*), Lake Eyre Hardyhead (*Craterocephalus eyresii*), Spangled Grunter (*Leiopotherapon unicolor*), Central Australian Catfish (*Neosilurus argenteus*), Rainbow Fish (*Melanotaenia splendida*), Central Australian Goby (*Chlamydogobius eremius)* and Western Chanda Perch (*Ambassis castelnaui*).

Lake Eyre itself is a barrier to such widespread dispersal in other fish. In the absence of 'rains of fishes' (see box), dispersal between river systems (or within a river system for that matter) occurs through floods. Widespread transport usually means going through the ultimate terminus at Lake Eyre. Species unable to cope with the temporary and predominantly highly saline waters of Lake Eyre will be unable to populate other river systems.

Since the beginning of European settlement, Lake Eyre has filled only a few times. In 1976 it was at its greatest depth since European settlement. From the level of ancient beaches deeper floods have occurred on only three occasions; at 500, 1000 and 3000 years previously. During these historic events all forms of wildlife explode into a frenzy of activity. Tens of millions of fish flood into the lake, which at first has correspondingly low salinity levels. But as the water evaporates, salinity levels increase and periodic fish kills of different species occur one at a time. In 1976 for instance, four strandlines, about one metre wide and about 25 to 30 metres apart, each comprising different fish, were found around the drying lake. In the highest strand were those fish which could barely tolerate salinity and the lowest strand those fish which were virtually pickled alive before succumbing.

Apart from salinity, other chemical and thermal barriers do not seem to limit the overall distribution of inland fish. Nevertheless, most of the species show some preference for specific habitats. Bony Bream has a broad habitat preference and is found in high numbers in the middle of Coongie Lake for instance, but other fish prefer sheltered, vegetated habitats such as swamps and backwaters. Within these habitats there is also a complex suite of fish movements involving vertical, lateral, longitudinal and diurnal migrations.

The aquatic channel and lake habitat of Coongie Lakes also supports the richest frog community in the arid zone. Summer rain initiates a deafening chorus of frogs. About eight species of water-holding frogs, tree-frogs and grass-frogs have their refuge here. Adjacent floodplains and ephemeral pools are especially important for the frogs. Populations of the water-rat (*Hydromys chrysogaster*), and an endemic species of tortoise (*Emydura* sp.) feed on a rich supply of invertebrates as well as fish and frogs. The water-rat, being a rodent, is unusual in that it rarely eats vegetation. But it will eat lizards, small mammals and even waterbirds.

Left, a flood of water into Coongie Lakes sparks off a flush of growth. Below, with an evaporation rate of around 3.5 metres per year, even Coongie Lake itself can dry out in less than a year.

THE GOBY

The small Central Australian Goby (*Chlamydogobius eremius*), is endemic to arid Australia. Its diverse ecological adaptation is a good example of what is necessary to survive in its environment. It occupies all types of water from very fresh (even distilled) to salinities higher than sea water. It can tolerate temperatures as low as five degrees Celsius and as high as 40 degrees Celsius. Behaviourally it uses the layered temperatures in a pool, congregating in cooler vegetated side-shallows or lying buried in the cooler bottom sediments. It will even emerge to allow itself to cool evaporatively. It can survive in almost stagnant water and will use the respiratory refuge provided by photosynthesising algal mats which bubble oxygen into surrounding water. As if this was not enough, it can also tolerate salinities from 0.2 to 37 parts per thousand – the equivalent of sea water.

The riparian woodlands are important nesting sites for the water-rat and tortoise. The woodland also supports a distinctive reptile community. Four species of reptiles — two skinks, a legless lizard and a dragon — are confined to the denser woodlands associated with channels and lake margins. Others are confined to floodplain habitat.

The reptile fauna, predictably, is not as rich as that in the Simpson Desert to the north. Furthermore, the two regions, while adjacent, have quite different suites of reptiles. More specifically, the Cooper region has fewer skinks and varanids but more snakes which prey on mammals and nesting birds found more regularly in the floodplains.

Mammals are abundant in the Coongie Lakes region, with about 12 native species including two planigales, *Planigale tenuirostris* and the larger *Planigale gilesi*. These uncommon carnivorous marsupials prefer deep cracking clays and are found in floodplain, lake-shore and channel-edge habitats and in deeply cracked claypans. The planigales typically have flattened, triangular heads which are easier to squeeze into the narrow cracks in which they shelter from summer heat and winter cold.

The rarer Forrest's Mouse (*Leggadina forresti*) is found on floodplains with a cover of low chenopods and at the margin of salt lakes and densely vegetated ephemeral lake-beds.

The Long-haired or Plague Rat (*Rattus villosissimus*) is usually rare over the continent as a whole. But after wet periods it can plague in great numbers. Between plagues, the species distribution again shrinks to small pockets in perennially damp locations within the Channel Country. The Coongie Lakes with its dense lignum beds adjacent to lakes and channels is probably such a refuge.

Perhaps the most stunning and globally significant aspect of Coongie Lakes is its concentration of birds. In all there are about 205 species of water and terrestrial birds in the region — a higher species diversity than any other truly arid region in Australia.

At least 20 000 waterfowl, increasing to a phenomenal 35 000, occupy the Coongie Lakes all year. Grey Teal and Pink-eared Duck constitute the great majority of these. Other birds occurring in thousands are Hardhead, Maned Duck, Eurasian Coot, Red-necked Avocet and Pelicans, the latter descending on lakes in a spectacular spiral of a thousand birds.

Other waterbirds occurring in large concentrations are Pacific Black Duck, the rare Australasian Shoveler, Black-winged Stilt, Hoary-headed Grebe, Pied Cormorants, terns and Silver Gulls. Herons, ibis, spoonbills and a variety of waders also occur.

Many endangered species find refuge in the Coongie Lakes region. The endangered Bush Thick-knee, for instance, is found in floodplain habitats where it feeds in grasslands. About 1000 of the vulnerable Freckled Duck prefer the flooded lignum margins.

Also significant are the 17 species of raptors including the rare Grey Falcon, Letter-winged Kite and Black-breasted Buzzard. The Flock Bronzewings, which have experienced an ominous decline in numbers over the past 100 years, are found in hundreds in the Coongie Lakes district feeding on dense swards of Sandhill Spurge (*Phyllanthus*), a characteristic plant of the district's white dunes.

At least 100 other species of terrestrial birds are found in the Coongie Lakes region. These birds are largely dependent on the readily available drinking water and the River Red Gum and Coolabah (*Eucalyptus camaldulensis* and *Eucalyptus microtheca*) riparian woodland.

Spangled Grunters (Perch) (Leiopotherapon unicolor) *are one of the most widespread of inland fishes in Australia, found in every inland river in the Lake Eyre Basin, artesian springs, bores, dams and reservoirs.*

Seventeen species of raptors are found in their hundreds in the Coongie Lakes region. These birds include the rare Grey Falcon, Letter-winged Kite and Black-breasted Buzzard.

PLAGUES

Native Plague Rat (Rattus villosissimus).

'The rats were a great nuisance. Cattle and horses had hooves nibbled, saddlery was eaten, and people camped away from settlements were bitten in their sleep. To walk about at night with a torch was to see rats every few paces scurrying along worn trails to numerous holes, to see them foraging on the open grassland, and to hear them squeaking and fighting in every quarter. There was no other word than fantastic to describe it all. The scene was reminiscent of a rabbit plague in miniature.' (A. Newsome and L. Corbett, 1975.)

The Long-haired Rat or Plague Rat, *Rattus villosissimus*, is usually rare. But occasionally numbers rise from great rarity to great abundance in barely a year. The year 1969 produced a plague rated as one of the greatest on record. Following big rains in the summer of 1967 and a cyclone in 1968, the cracking clay plains of the Barkly Tableland, which is a refuge for the rat, turned into a sticky swamp. A prolonged super-abundance of grassy resources provided the trigger which sent invading waves of Plague Rat down the rivers of the Channel Country towards Lake Eyre.

The Plague Rats suddenly appeared in places where they had never been seen before; in Katherine in the tropics, on the border of Western Australia, in central Australia in the Tanami Desert and in the Simpson Desert.

The mammalian bounty provided by the rodent plagues attracted predators. The normally rare and nocturnal Letter-winged Kite, which preys almost exclusively on the rat, became abundant. So did the Barn Owl and the Dingo and many introduced predators (180 cats were shot around Uluru). It was these animals that finally put the lid on the plague, even before the run of good seasons ended.

A lot of attention has been paid to animals which plague in the Australian outback. So much so that plaguing is seen to be a major adaptation of desert dwellers. It is important to remember, however, that of the original fauna that was present only a minority would have been plaguing forms.

Another animal which irrupts under favourable conditions is the plague locust (*Chortoicetes terminifera*). It, too, is an animal of the Channel Country and is dependent on the rich resources of the tussock grasslands.

During drought when numbers of locusts are low, they behave as individuals. When numbers are high, as a response to drought-breaking rain and green growth, they switch to a different behavioural phase and act as highly mobile, gregarious units. As well as rain, which brings on green vegetation and hatches drought-quiescent eggs, upper air patterns allowing locusts to migrate with the strong winds are crucial for a fully-fledged plague.

Lake Eyre lies in the heart of the Australian continent. Its lowest point lies 15.2 metres below sea level. Lake Eyre north is 144 kilometres long and 77 kilometres wide. It is joined to Lake Eyre south, which is 64 kilometres long and 24 kilometres wide, by the narrow Goyder Channel. The combined area is 9690 square kilometres.

River red gums and Coolabahs line the rivers of sand right across the arid zone. These trees are analogous to a multi-story, high density housing development for birds. Along the Cooper, for instance, the fringing woodland has a dense crown cover and at least four well-developed levels. Species mainly confined to this habitat on the Cooper are the vocal Barking Owl, which spreads out evenly in pairs along the Creek, Mallee Ringneck Parrot, Sacred Kingfisher and Restless Flycatcher. A colourful and rich variety of other songbirds are also associated with the woodland habitat.

Interestingly, a number of birds which prefer a more arid habitat do not occur in these riverine woodlands. Birds such as the Inland Thornbill, Pink Cockatoo, Hooded Robin, Rufous Whistler and others have distributions around the region but not in or through it.

This tortoise, Emydura sp., may well be endemic to the region. Work continues on its taxonomy.

The floodplain with its 1–2 metre high shrublands dominated by Lignum (*Muehlenbeckia cunninghamii*) also has a characteristic suite of birds including the Chirruping Wedgebill and the Banded Whiteface.

This is also the habitat of the rare Grey Grasswren. Not discovered until 1967, the Grey is the only grasswren living in swampland of any sort. It prefers dense lignum and swamp cane grass of at least one metre in height. These plants may be needed as much for their height — above the floods which spread out every few years — as for the shelter they provide. The birds eat seeds and insects and, at times, water-snails. They have a strangely restricted distribution, so far only at Goyder's Lagoon, at the mouth of the Diamantina, and in a similar habitat at the mouth of Bullo River. The birds appear to be residual colonies of a once widespread population which would have ranged over a greatly expanded swampy region prior to the last glacial.

Yellow-billed Spoonbills near Coongie Lake.

TUSSOCK GRASSLANDS

The Barkly Tableland forms part of what is known as the tussock grasslands. These arid grasslands of Mitchell Grass, predominantly *Astrebla* species, cover 500 000 square kilometres in an arc mainly on the eastern and north-eastern margin of the desert. The productive grasses grow on rich cracking clay plains; essentially fine, fertile mud deposited by rivers over millions of years.

When wet the soil expands and the plains turn into a swampy quagmire. But during winter, when it is dry, the soil shrinks into rough columns and gaping cracks appear. These cracks rip any roots growing across them. This probably explains why no trees grow on the plains. The Mitchell Grass has long, vertical roots which are left unbroken and only pushed to one side or the other when the soils crack. As many as 50 species of native grass grow here even in winter.

Bustard.

The rich supply of seed and productive grass provides a reliable resource for grasshoppers and many seed-eating birds such as Button-quails, Cockatiels and Budgerigars. Also present are Bustards avidly feeding on grasshoppers, beetles, fruits and seeds. This most goofy-looking bird assumes a frozen position when disturbed in the hope that it will not be noticed. After a while it tentatively stalks off. The tussock grasslands are also the last stronghold for hundreds of seed-eating Flock Bronzewings; and also for venomous snakes.

Snakes occur widely over Australia's deserts, but not in huge numbers. Most of them are harmless. The tussock grasslands are an exception. Venomous snakes are common here and some, like the Speckled Brown Snake, are endemic to the region. The snakes feast on the abundant mammal fauna which these fertile grasslands support. The cracks, into which snakes happily slither, provide ready-made refuges for the Plague Rat, planigales and other carnivorous marsupials, and also bats.

There are many surprises in this intriguing food chain. Predictably, lizards are not common in a foodweb based on seeds and green vegetation. Spencer's Monitor, however, likes living dangerously and preys on diurnal snakes such as Death Adders. When hunting, the goanna cleverly avoids the first strike and grabs the snake while it is off balance before the next strike.

Pools of Hidden Life

Another striking feature of the plain and sandhills of the low Lake Eyre Basin is the number of claypans and salt lakes. These vary from a few metres to a kilometre or more. Mostly they are bone dry with a thin surface film broken up into curled glistening flakes or where the mud is thick, deep fissures run down between roughly hexagonal masses of hardened earth.

One of the most primitive of living crustaceans is the Shield Shrimp (below). They seem to explode into life after good rains, turning claypans and muddy floodwaters into a living shrimp soup. Not only carapaces are left (above) when the pools dry out. Desiccated dust-sized eggs are left to blow about. They settle in unusual places, such as the top of Uluru (right), where Shield Shrimps have been found in ephemeral pools. Left, a close-up of the head of a Shield Shrimp.

The dead shells of molluscs, the carapaces of crustaceans and, where the water is fresh, tiny frog footprints show that they do at times contain abundant life.

Composed mainly of invertebrates, this cryptic life thrives within the murky waters of the newly-filled lakes. In particular, the salt lakes are inhabited by an assemblage of animals unique to the continent.

The most conspicuous invertebrates are the crustaceans. These are mostly endemic and demonstrate a surprising tolerance to high salinities, high temperatures and desiccation. The ostracods, seed-shrimps or shell-shrimps, are the most diverse group in Australia with 37 species, which is a much larger number compared with other countries. These small animals from 1.25 millimetres to 6 millimetres long are enclosed in a bivalved shell which gapes open to allow their legs to poke through. Ostracods can be swimmers, clingers, climbers or burrowers and their eggs can withstand complete desiccation. Two species of amphipods, otherwise called freshwater shrimps or 'side-swimmers', are also common in Australian salt lakes. The name 'side-swimmer' comes about because the animal is stongly laterally compressed, in other words it looks squashed-in from the sides.

Perhaps the best known of invertebrates, however, are the Anostracan brine- or fairy-shrimps which are also endemic to Australia with nine species. These delicate

Left, a short but furious existence ends for these Shield Shrimps.

Below, freshwater shrimps can walk backwards or forwards or may swim with the aid of little swimmerets.

The thumbnail-sized fairy-shrimps swim legs uppermost. They can tolerate water ten times saltier than sea water. Like Shield Shrimps, their eggs are resistant to desiccation.

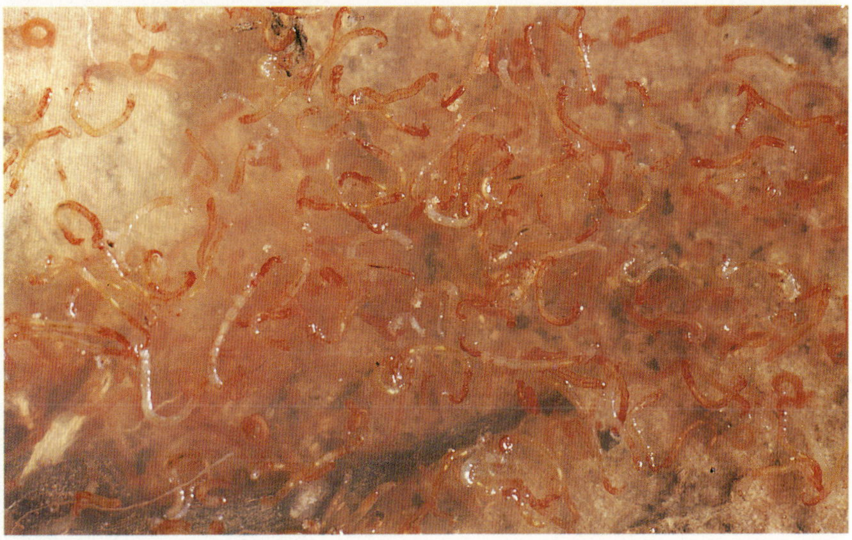

Mosquito and midge larvae can breed in just a few millimetres of water and in just a few days. Midge larvae are some of the most tolerant of high salinity.

animals are about the size of a thumbnail, and with the shield shrimps, are amongst the most primitive of all living crustaceans. Quite incredibly they can tolerate salinities of 353 parts per thousand, which is ten times saltier than sea water and is almost at salt saturation point.

The fairy-shrimps swim with their legs uppermost, apparently upside down. They feed on micro-organisms collected in a groove on the thorax. Their eggs are resistant to desiccation and may even require a period of desiccation for development to take place.

Gastropods or snails are other significant components of salt lake fauna.

More obvious, perhaps, are the insects. Salt water is an adverse environment for most insects. Whether they occur or not is detemined mainly by the permanence of the water and the salinity. A variety of novel strategies have evolved to cope with the trying conditions. Some dragonfly nymphs, for instance, can survive in damp soil for months whereas others breed in large numbers and complete their life cycles in 36 days.

Mosquitoes appear to have it all sewn up. They can breed in just a few millimetres of water and in a period of days. They can survive very high water temperatures and salinities and the larvae eat bacteria which begin multiplying immediately in ephemeral waters. Not many other animals can survive the conditions, so predators and competitors are few and mosquito larvae can quickly build up.

When the pools foul so that it is impossible to respire, some mosquitoes pierce vascular plant tissue where they obtain air. Other aquatic insects swim to the surface for air, capture a bubble of oxygen from algae or have long breathing tubes. The most tolerant insects of high salinity are mosquitoes, midges and gnats, caddis flies and a variety of water beetles.

Above, the common yabbie (Cherax destructa) *varies in colour from murky green through pale blue to brighter blue. It is believed the bluer yabbies inhabit clearer waters. Left, a sheild shrimp laying eggs in a muddy pool in the Tanami Desert. Above right, a shield shrimp in the Great Victoria Desert. Right, these Chironomid larvae (midges) contain haemoglobin in their blood, allowing them to extract the small quantity of oxygen available in stagnant pools.*

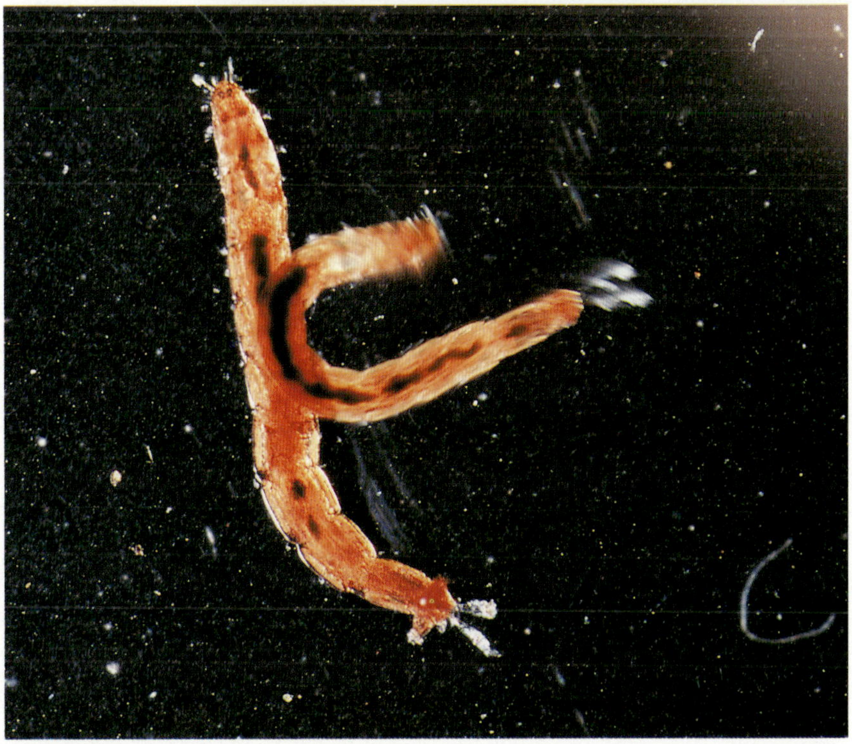

RAINS OF FISHES

Preposterous! That fishes should fly across the sky seems the height of absurdity. Curiously enough though, it does indeed seem possible that fishes can be transported through the air and deposited afar during storms. There have been many otherwise inexplicable examples of this occurrence. These include a report of small fishes rattling on a tin roof during a storm in Caloundra, Queensland in 1909; hundreds of gudgeons in the streets in Gulargambone, New South Wales around 1920; small fishes in ponds after a severe cyclone in Marble Bar in Western Australia in 1942; and numerous grunters about 5 centimetres long on the airstrip between 3–6 a.m. at Hughenden in Queensland in 1971.

When the sun came out, after days of rain, it was too hot, even for flies. Storm clouds lined the northern horizon, with dark lines of rain trailing from them. It was the Tanami Desert in summer. A dull roar, like a deep-seated earthquake, followed by a gurgling and rushing sound announced the arrival of a flash flood. Rarely seen on a sandplain, it was an amazing sight. The water used a compacted track to push ahead froth and debris, making way for the water to come. Minutes later, the track was a creek, knee-deep, and the next day the water was gone.

Driving from Alice Springs to Halls Creek through the Tanami Desert, the palaeorivers, while obvious from space, can only be detected on the ground by a change in termite species. Compared to the diminutive spires built by others, this particular species builds a mound slapped together like an over-filled ice-cream cone. A closer look at the environment will reveal *Melaleuca* and greener spinifex, hinting at a concentration of nutrients and water. Many of the small and medium-sized mammals which inhabit what seems like hostile (to herbivores) spinifex grasslands are concentrated on buried drainage lines which form deceptive ecological refuges. What may appear counter-intuitive to us was recognised by the Aborigines for millennia because it is here that Aborigines concentrate their fire-stick farming practices, encouraging herbivores such as bandicoots and wallabies and also catalysing the germination of specific food plants: bush tucker.

The Lake Amadeus Salt Lake System is one palaeodrainage system which is far from cryptic. It dwarfs Uluru (Ayers Rock) and Kata Tjuata (the Olgas) and has long been recognised as a major linear feature on the topographic map of Australia. The system is a 500 kilometre long chain of playas from Lake Hopkins in Western Australia, through Lake Neale and Lake Amadeus to the Finke River in the Northern Territory. It drains a colossal catchment of 90 000 square kilometres. Yet this great valley is devoid of surface drainage. Several intermittent streams drain the Petermann and Musgrave Ranges to the south and the Cleland Hills and George Gill Ranges to the north. All of these streams, however, flood out in sand dunes.

The Lake Amadeus system is remarkable even in Australia for its low regional relief and negligible surface run-off. Salt accumulation is by slow evaporation of groundwater seeping to the surface and so the Amadeus system does not have the thick salt crusts of other salt lakes such as Lake Eyre and Lake Frome. The total evaporative playa surface of the Lake Amadeus system is 1750 square kilometres. Quaternary sand dunes up to 30 metres high cover a large part of the surface depression. And it is this system which gives us evidence of the earliest aridity, as we know it, in the Australian continent.

Mound Springs

More than anything else desert environments are driven by water. And in this book we have traced the flow of water through Australia's deceptively uniform desert environment; from the desert mountains with their juxtaposition of shapes; through the gently sloping mulga woodlands and the spinifex grasslands; down into the wierdly beautiful breakaway country and gibber plains; and finally into Lake Eyre, the lowest part of the continent.

Underlying the surface waters of the Lake Eyre Basin are the ancient waters of the Great Artesian Basin. Water from rain falling along the western slopes of the Great Dividing Range in Queensland and New South Wales seeps into the aquifers of this 1 700 000 square kilometre basin. In turn, water is lost when it leaks into other watertables or into neighbouring basins, or bubbles and gushes out on the surface in an arc of mound springs west of Lake Eyre.

These mound springs are extraordinary oases, seeping water that fell as rain perhaps two million years ago. They are the last desert refuges we will explore.

The largest concentration of springs are in the southern, south-central and south-western parts of the basin. The areas south-west and north-west of Lake Eyre contain the most active springs, the largest of these being the Dalhousie Springs, a region of around 70 square kilometres, 250 kilometres north-west of Lake Eyre, containing about 80 active springs. The Dalhousie Springs complex discharges 43 per cent of the natural discharge of water from the Great Artesian Basin; enough to maintain small creeks for some kilometres.

Dalhousie Mound Springs contain deep, bathtub-warm water. Surrounded by a curtain of vegetation, it is the ideal place to wash off the desert's dust.

Dalhousie Springs proper contains large (50 metres long, 10 metres deep) inviting bathtub pools of warm water surrounded by a thick curtain of vegetation. Of the 100 or so species of terrestrial and semi-aquatic plants associated with the spring, the most conspicuous are the tea-tree (*Melaleuca glomerata*), which here grows to 10–12 metres, the common reed (*Phragmites australis*) and the bulrush (*Typha domingensis*). Only one plant, the native tobacco, *Nicotianà burbidgeae*, is endemic to the springs.

In contrast, the fauna of the Dalhousie Springs complex has many endemic species. The fish fauna in particular is noted for its unique forms. Three of the six fishes occurring at Dalhousie Springs are endemic. These are the undescribed Dalhousie Catfish (*Neosilurus* sp. nov.), the Dalhousie Hardyhead (*Craterocephalus*

FORMATION OF MOUND SPRINGS

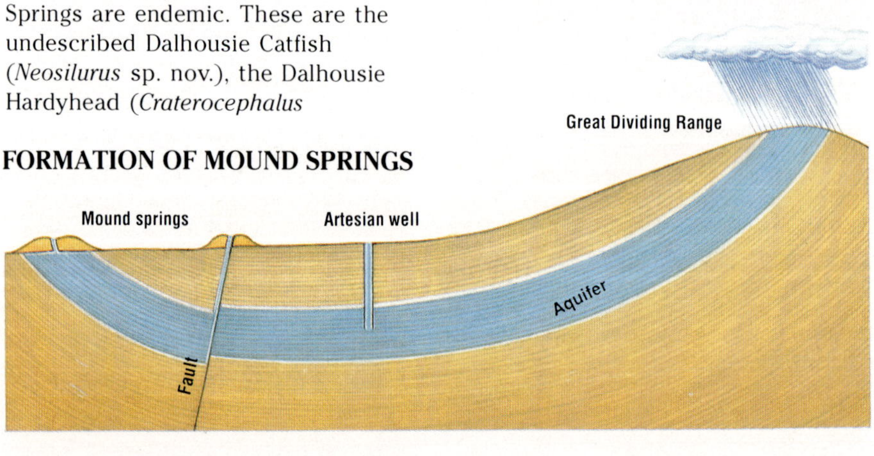

Water from rain seeps into a vast, porous layer of rock, or aquifer, along the continent's Dividing Range. The water filters through the aquifer and bubbles out where the aquifer reaches the surface thousands of kilometres away or where a fault railroads the water upwards.

Dalhousie Springs was an important area for the Southern Arrernte people and their neighbours, the Wangkangurru of the Simpson Desert. It is associated with the Perentie (goanna) ancestor of the Dreamtime.

Dalhousie hardyhead (Craterocephalus dalhousiensis) *are restricted to Dalhousie Spring. The fish dart briefly into the central hot parts of the pool, which are up to 43° C, probably to feed on the blue green bacteria that grow there.*

dalhousiensis), and the undescribed Dalhousie Goby (*Chlamydogobius* sp. nov.) Other possible endemic species include at least six species of snail, several small crustaceans, a yabbie and a frog.

Clearly, Dalhousie Springs must have been isolated from other freshwater systems for some time for this extraordinary concentration of unique species to have evolved.

Dalhousie Springs is within the Finke River catchment. In fact it marks the end of the ancient river. As far as anyone can tell, the Finke has not flowed at the latitude of Dalhousie Springs for about 10 000 years and is now choked by its own sand. This isolation has allowed the evolution of Dalhousie Springs' unique species to occur.

In many ways this evocative region provides a fitting ending. It is where the Finke, ancient long before mound springs ever erupted, spills its drought-breaking water into a sea of sand and where it is possible to wash off the desert dust in spring water which originally fell as rain 1500 kilometres away during moister climates of the past when the *Diprotodon* roamed the chenopod shrublands and came to the river and lake shores to drink, watching warily for prowling *Thylacoleo*.

River of sand, north of the Petermann Ranges.

NUISANCE VALUE OF BUSH FLIES

Why is it that the smallest of bush flies drive us the most crazy? They crawl into the eyes, ears, nose and wait for any opportunity to buzz down the throat. As it happens, there is a perfectly good biological reason for this behaviour.

The common bush fly, *Musca vetustissima*, occurs right across the continent. But, more than any other region, it is a creature of the wide open spaces. In fact, except for roosting in the evening and avoiding searing temperatures, it does not like shade. This explains why it does not like forests.

What it does like, however, is a moist pat of well-aerated dung. In dung, small creamy-white eggs are carefully deposited. The larvae, or maggots, that hatch from these eat their way through the dung. After 60–120 hours they leave the home pat and bury themselves in the soil to pupate; the longer the maggot has had to feed, the bigger the resulting fly will be. If the temperature is warm, a grey fly will emerge from the pupa after only three days. After three more days the flies are ready to mate, the male 'jumping' the female in flight. Copulation lasts around 80 minutes.

The adult, with mouthparts that consist of a lapping-type proboscis with rasping teeth, needs a diet of protein; blood, pus, milk, tears and saliva are best — sweat and urine seem to be inadequate protein sources but are good for moisture. Dung is a poor source of protein. The smaller the fly the more starved it is of protein. The situation is exacerbated in females who require extra protein for egg development. In their insatiable quest for protein, these become the most irritating and tenacious creatures, outnumbering their brothers three to one in the search for your essential juices.

In fact, the 'nuisance value' of flies has actually been measured. If we put our hand in a cage of flies, it can be shown that around 10 blood-fed flies will be attracted to your hand per minute, whereas the figure increases to 48 flies per minute when they are protein starved. Generally large blood-fed flies have the smallest nuisance value. Flies which had an underprivileged upbringing in a lower-class dung heap and, once adult, continued to be undernourished, are up to 70 times a greater nuisance.

Flies also stick to big animals like us knowing that we are eventually going to drop some dung. Fresh dung is particularly attractive to gravid females. Moreover, bushflies are stongly attracted upwind by the odour of freshly-dropped dung. After munching away at the dung for a while the females scurry over the dung, exploring all its nooks and crannies for a suitable site for the eggs. Once one female starts laying it attracts others, perhaps by some sort of pheromone, and they gather in groups to lay.

Bush flies are found everywhere in Australia, particularly in the geographical centre of the continent. It is enough to lose your head!

BIBLIOGRAPHY

Archer, Michael 1982, *Carnivorous Marsupials*, vol. 1, Royal Zoological Society of New South Wales, Sydney

Baker, V.R., Pickup, G., & Polach, H.A. 1983, 'Desert palaeofloods in central Australia', *Nature*, vol. 301, issue 5900, pp. 502–4

Barker, W.R. & Greenslade, P.J. M. 1982, *Evolution of the Flora and Fauna of Arid Australia*, Peacock, Adelaide

Barlow, B.A. 1971, 'Cytogeography of the Genus *Eremophila*', *Aust. J. Bot*, vol. 19, pp. 295–310

Barlow, B.A. 1981, 'The Australian flora: its origin and evolution', in *Flora of Australia*, Bureau of Flora and Fauna, AGPS, Canberra

Beadle, N.C.W. 1966, 'Soil phosphate and its role in molding segments of the Australian flora and vegetation, with special reference to xeromorphy and sclerophylly', *Ecology*, vol. 47, no. 6pp. 992–1007

Beard, J.S. 1976, 'The evolution of Australian desert plants' in Goodall, D.W. (ed) *Evolution of Desert Biota*, University of Texas Press, Austin, U.S.A.

Bolam, 1927, in Copley, P. 1988, 'The Stick-nest Rats of Australia', a final report to the World Wildlife Fund (Australia), Department of Environment and Planning South Australia, D.J. Woolman, Government Printer, Adelaide

Bowler, J.M. & Wasson, R.J. 1984, 'Glacial Age environments of inland Australia' in Vogel, J.C.(ed) *Late Cainozoic Palaeoclimates of the Southern Hemisphere*, A.A. Balkema, Rotterham

Bryceson, K.P. & Wright, D.E. 1986, 'An analysis of the 1984 locust plague in Australia using multitemporal Landsat multispectoral data and a simulation model of locust development', *Agriculture, Ecosystems and Environment*, vol. 16, pp. 87–102

Burbidge, Andrew A. & McKenzie, N.L. (in press) 'Patterns in the modern decline of Western Australia's vertebrate fauna: causes and conservation implications', *Biol. Cons.*, vol. 50, pp. 143–198

Caughley, Graeme, Chepherd, Neil & Short, Jeff 1987, *Kangaroos: their ecology and management in the sheep rangelands of Australia*, Cambridge University Press, Cambridge, U.K.

Chen, Xiang-Yang 1989, 'Lake Amadeus, Central Australia: modern processes and evolution', unpublished PhD thesis, Australian National University, Canberra

Cogger, Harold G. 1991, *Reptiles and amphibians of Australia*, Reed Books, Sydney

Copley, P. 1988, 'The Stick-nest Rats of Australia', Department of Environment and Planning, South Australia, D.J. Woolman, Government Printer, Adelaide

Corbett, L.K. 1985, 'Morphological comparisons of Australian and Thai dingoes: a reappraisal of dingo status, distribution and ancestry', *Proc. Ecol. Soc. Aust.*, vol. 13, pp. 277–91

Davidson, D.W. & Morton, S.R. 1981, 'Myrmecochory in some plants (*F. chenopodiaceae)* of the Australian arid zone', *Oecologia*, vol. 50 pp. 357–66

Davidson, D.W. & Morton, S.R. 1984, 'Dispersal adaptations of some *Acacia* species in the Australian arid zone', *Ecology*, vol. 65, no. 4, pp. 1038–51

Davies, S.J.J.F. 1986, 'A biology of the desert fringe', *R. Soc. of W.A.*, vol. 68, no. 2, pp. 37–50

De Decker, Patrick 1983, 'Australian salt lakes: their history, chemistry, and biota — a review', *Hydrobiologie* 105, pp. 231-44

Denny, M.J.S. 1975 'Mammals of Sturt National Park, Tibooburra, New South Wales', *Aust. Zool.*, vol. 18, no. 3, pp. 179–95

Department of Primary Production Division of Agriculture and Stock, 1981, 'Botanical notes from the herbaria of the Northern Territory', *Northern Territory Botanical Bulletin, no. 4*

Dunlop, J.N. & Sawle, Maryanne 1982, 'The habit and life history of the Pilbara Ningaui *Ningaui timealeyi*', *Rec. West. Aust. Mus.* vol. 10, no. 1, pp. 47-52

Evans, P.R. 1988, 'The formation of petroleum and geological history of Australia' in *Petroleum in Australia: The first century*, APEA, Australian Petroleum Exploration Assoc. Ltd., Sydney

Favenc, Ernest 1983, *The history of Australian exploration 1788–1888*, Golden Press Facsimile Edition, Golden Press, Sydney

Gans, C., Merlin, R. and Blumer W.F.C. 1982, 'The Water-collecting mechanism of *Moloch horridus* re-examined' *Amphibia-Reptilia 3*, pp. 57–64

Graetz, R.D. & Howes, K.M.W. 1979, *Studies of the Australian arid zone 4, chenopod shrublands*, CSIRO, Melbourne

Greenslade, P.J.M. 1979, *A guide to ants of South Australia*, South Australian Museum, Adelaide

Greenslade, P.J.M. & Greenslade, Penny 1983, 'Ecology of Soil Invertebrates in Soils — an Australian Viewpoint.' *CSIRO*, Melbourne Academic Press, London

Greer, Allen E. 1979, 'A New Species of *Lerista lacertilia scincidas* from Northern Queensland, with remarks on the origin of the genus', *Rec. of the Aust. Mus*, vol. 32, no. 10, pp. 383-8

Greer, Allen E. 1987, 'Limb reduction in the lizard genus *Lerista*: 1. Variation in the number of phalanges and presacral vertebrae', *Journal of Herpetology*, vol 21, no. 4, pp. 267-76

Greer, Allen E. & Cogger, Harold G. 1985, 'Systematics of the reduce-limbed and limbless skinks currently assigned to the genus *Anomalapus* (Lacertilia: Scincidae)', *Rec. of the Aust. Mus.*, vol. 37, no. 1, pp. 11–54

Griffin, G.F. (in press) 'Characters of three spinifex communities in central Australia', *Journal of Vegetation Science*, CSIRO, Alice Springs

Griffin, G.F. (in prep) 'Changes in vegetation and soil chemistry after fire in three spinifex communities' CSIRO, Alice Springs

Gullan, Penny J. & Cockburn, Andrew 1986, 'Sexual dichronism and intersexual phoresy in gall-forming coccoids', *Oecologia*, vol. 68, pp. 632–4

Harrington, G.N., Wilson, A.D. & Young, M.D. 1984, *Management of Australia's Rangelands*, CSIRO, Melbourne

Hnatiuk, R.J. & Maslin, B.R. 1988, 'Phytogeography of *Acacia* in Australia in relation to climate and species-richness', *Australian Journal of Botany*, vol. 36, no. 4, pp. 361–83

Hobbs, T.J. 1987, *The breakaways reserve; a resource inventory*, Department of Lands, Adelaide

Howes, K.M.W. 1978, *Studies of the Australian arid zone III: Water in rangelands*, CSIRO, Melbourne

Hubble, G.D. and Isbell, R.F. 1983, *Soils: An Australian Viewpoint.* CSIRO, Melbourne; Academic Press, London

Hughes, R.D. & Nicholas, W.L. 1972, 'A synopsis of observations on the biology of the Australian bushfly, *Musca vetustissima* Walker', J. *Aust. Ent. Soc.*, vol. 11, pp. 311–31

Jacobson, G. 1988, 'Hydrology of Lake Amadeus, a groundwater-discharge playa in central Australia', *BMR Journal of Australian Geology and Geophysics*, vol. 10, pp. 301–8

Jacobson, G., Arakel, A.V. & Chen, Yijian 1988, 'The central Australian groundwater discharge zone: evolution of associated calcrete and gypcrete deposits', *Australian Journal of Earth Sciences*, vol. 35, pp. 549–65

Keast, Allan 1959, 'Relict animals and plants of the MacDonnell Ranges', *Aust. Mus. Mag.*, vol. 13, pp. 81–86

Latz, P.K. 1975, 'Notes on the relict palm *Livistona mariae* F. Muell. in Central Australia', *Trans. R. Soc. of S.A.*, vol. 99, no. 4, pp. 189–96

Latz, P.K., Johnson, K.A. & Gillam, M.W. 1981, 'A biological survey of the Kings Canyon area of the George Gill Ranges', Internal Report, Conservation Commission of the Northern Territory, Alice Springs

Loffler, E & Sullivan, M.E. 1979, 'Lake Dieri resurrected: an interpretation using satellite imagery', *Z. Geomorph. N.F.*, vol. 23, no. 3, pp. 233–42

Low, W.A. et al 1978, 'The physical and biological features of Kunoth Paddock in central Australia', *Division of Land Resources Management Technical Paper No 4*, CSIRO, Australia

Lundie-Jenkins, G.W. 1989, 'The ecology and management of the Rufous Hare-wallaby *Lagorchestes hirsutus* in the Tanami Desert', (unpublished report) Conservation Commission of the Northern Territory, Alice Springs

Maclean, G.L. 1976, 'A field study of the Australian pratincole', *Emu*, vol. 76, no. 4, pp. 171–83

Maclean, G.L. 1976, 'A Field Study of the Australian Dotterel', *Emu*, vol. 76, no. 4, pp. 207–16

MacMillen, R.E. & Lee, A.K. 1969, 'Water metabolism of Australian hopping mice', *Comp. Biochem. Physiol.*, vol. 28, pp. 493–514

MacMillen, R.E. & Lee, A.K. 1970, 'Metabolism and pulmocutaneous water loss of Australian hopping mice', *Comp. Biochem. Physiol.* vol. 35, pp. 355--69

Main A.R. 1976, 'Adaptation of Australian vertebrates to desert conditions' in Goodall, D.W. (ed) *Evolution of Desert Biota*, University of Texas Press, Austin

Maslin, B.R. 1982, 'Studies in the genus *Acacia* (Leguminosae: Mimosoideae)-11. *Acacia* species of the Hamersley Range area, Western Australia.' *Nuytsia*, vol. 4, no. 1, pp. 61-103

Matthews, E.G. & Kitching, R.L. 1984, *Insect Ecology*, 2nd edn, University of Queensland Press, Brisbane

McKenzie, N.L. & Robinson, A.C. 1987, *A biological survey of the Nullarbor region south and Western Australia in 1984*, South Australian Department of Environment and Planning, Western Australian Department of Conservation and Land Management and the Australian National Parks and Wildlife Service, Government Printer, Adelaide

Metting, Blaine 1981, 'The systematics and ecology of soil algae', *The Botanical Review*, vol. 47, pp. 195–312

Milnes, A.R. & Twidale, C.R. 1983, 'An overview of silicification in Cainozoic landscapes of arid central and southern Australia', *Aust. J. Soil Res*, vol. 21, pp. 387–410

Morton, S.R. 1982, 'Dasyurid marsupials of the Australian arid zone: an ecological review' in Archer, M. 1982, *Carnivorous Marsupials*, Roy. Zool. Soc. N.S.W., Sydney

Morton, S.R. & Baynes, A. 1985, 'Small mammal assemblages in arid Australia: a reappraisal', *Aust. Mammal*, vol. 8, pp. 159–69

Newsome A.E. & Corbett, L.K. 1975, 'Outbreaks of rodents in semi-arid and arid Australia: causes, preventions, and evolutionary considerations' in Prakash, I. & Gosh, P.K. (eds) *Rodents in Desert Environments*, Junk, The Hague

Pedley, L. 1973, 'Taxonomy of the *Acacia aneura* complex', *Tropical Grasslands*, vol. 7, no. 1, pp.3–8

Pianks, Eric R. 1972, 'Zoogeography and speciation of Australian desert lizards: an ecological perspective', *Copeia*, vol. 1, pp. 127–45

Pickup, G., Baker, V.R. and Allan, G. (1988) 'History, Palaeochannels and palaeofloods of the Finke River, Central Australia', in Warner, R.F. (ed) *Essays in Australian Fluvial Geomorphology*, Academic Press, Sydney

Preece, P.B. 1970, 'Contributions to the biology of mulga: 11. germination', *Aust. J. Bot*, vol. 19, pp. 39–49

Preece, P.B. 1970, 'Contributions to the biology of mulga: 1. Flowering', *Aust. J. Bot*, vol. 19, pp. 21–38

Pressland, A.J. 1973, 'Rainfall partitioning by an arid woodland *(Acacia aneura* F. Muell.) in south-western Queensland', *Aust. J. Bot*, vol. 21, pp. 235–45

Pressland, A.J. 1976, 'Soil moisture redistribution as affected by throughfall and stemflow in an arid zone shrub community', *Aust. J. Bot*, vol. 24, pp. 641–9

Purdie, Rosemary 1984, 'Land systems of the Simpson Desert region', *CSIRO Division of Water and Land Resources Natural Resources 2*, Sydney

Randell, B.R. & Symon, D.E. 1977, 'Distribution of *Cassia* and *Solanum* species in arid regions of Australia', *Search*, vol. 8, pp. 206–7

Reid, J.R.W., Badman, F.J. & Parker, S.A., in Tyler, M.J., Twidale C.R., Davies M., Wells C.B. *Natural History of the North East Desert*, Royal Society of South Australia Inc. 1990, 'Birds of the North Eastern Deserts of South Australia'

Reid, Julian & Gillen, Jake 1988, *The Coongie Lakes Study*, (unpublished report), Department of Environment and Planning, South Australia

Ritchie, Alexander & Gilbert-Tomlinson, Joyce 1977, 'First Ordovician vertebrates from the Southern Hemisphere', *Alcheringa*, vol. 1, pp. 351–68

Sattler, P.S. 1986, *The Mulga Lands*, Royal Society of Queensland, Brisbane

Schodde, Richard 1982, *The Fairy-wrens*, Landsdowne Editions, Sydney

Shorthouse, D.J. & Marpules, T.G. 1980, 'Observations on the burrow and associated behaviour of the arid zone scorpion *Urodacus yaschenkoi* (Birula)', *Aust. J. Zoo*, vol. 28, pp. 581–90

Singh, G. 1988, 'History of aridland vegetation and climate: a global perspective', *Biol. Rev*, vol. 63, pp. 159–95

Singh, G. 1981, 'Late Quaternary pollen records and seasonal palaeoclimates of Lake Frome, South Australia', *Hydrobiologia*, vol. 2, pp. 419–30

Smith, M.A. 1989, 'The case for a resident human population in the central Australian Ranges during full glacial aridity', *Archaeol. Oceania*, vol. 24, pp. 93–105

Spencer, Baldwin 1896, *Horn Scientific Expedition*, Melville, Mullen & Slade, Melbourne

Stewart, A.J., Blake, D.H. & Ollier, C.D. 1986, 'Cambrian river terraces and ridgetops in central Australia: oldest persisting landforms?', *Science*, vol. 233, pp. 758–61

Strahan, Ronald 1983, *The Complete Book of Australian Mammals*, Angus and Robertson, Sydney

Taylor, S.G. & Shurcliff, K.S. 1983, 'Ecology and management of central Australian mountain ranges', in Purdie, R.W. & Noble, I.R. (eds) *Mountain Ecology in the Australian Region*, Proceedings of the Ecological Society of Australia 12, Canberra

Tindale, N.B. 1953, 'On some Australian Cossidae including the moth of the witjuti (witchetty) grub', *Trans., Roy. Soc. S. Aust.*, vol. 76, pp. 56–65

Tindale, N.B. 1961, 'A new species of *Chlenias* (Lepidoptera, Boarmiidae) on *Acacia aneura*, with some central Australian native beliefs about it', *Records of the S.A. Museum*, vol. 14, pp. 131–40

Tolcher, H.M. 1986, *Drought or deluge: man in the Cooper Creek region*, Melbourne University Press, Melbourne

Thomson, B.G. 1990 *A Field Guide to the Bats of the Northern Territory*, N.T. Government Printer, N.T.

Twidale, C.R. & Bourne, Jennifer A. 1978, 'Bornhardts developed in sedimentary rocks, central Australia', *S.A. Geograaf*, vol. 6, no. 1, pp. 35–50

Twidale, C.R. & Milnes, A.R. 1983, 'Aspects of the distribution and disintegration of siliceous duricrusts in arid Australia', *Geologie en Mijnbouw*, vol. 62, pp. 373-82

Tyler, Michael J. 1989, *Australian Frogs*, Viking O'Neil, Adelaide

Tyndale-Biscoe, Marina 1989, 'The influence of adult size and protein diet on the human-oriented behaviour of the bush fly, *Musca vetustissima* Walker (Diptera: Muscidae)', *Bull. Ent. Res*, vol. 79, pp. 19–29

van de Graaff, W.J.E., Crowe, R.W.A., Buting, J.A. & Jackson, M.J. 1977, 'Relict Early Cainozoic drainages in arid Western Australia', *Z. Geomorph.*, vol. 21, no. 3, pp.379–400

Wassan, R.J. 1983, 'Dune sediment types, sand colour, sediment provenance and hydrology in the Strzelecki-Simpson dunefield, Australia', in Brookfield, M.E. & Ahlbrandt, T.S. (eds) *Eolian Sediments and Processes*, Elsevier, Amsterdam

Wassan, R.J. 1984, 'Late Quaternary palaeoenvironments in the desert dunefields of Australia' in Vogel, J.C. (ed) *Late Cainozoic Palaeoclimates of the Southern Hemisphere*, A.A. Balkema, Rotterdam

Wassan, R.J. 1986, 'Geomorphology and Quarternary history of the Australian continental dunefields', *Geographical Review of Japan 59* series B, vol. 1, pp. 55–67

Wassan, R.J. & Hyde, R. 1983, 'Factors determining desert dune type', *Nature*, no. 304, pp. 337– 9

Westoby, M., Rice, B., Griffin, G. & Friedel, M. 1988, 'The Soil Seed Bank of *Triodia basedowii* in relation to time since fire', *Aust. J. Ecol.*, vol. 13, pp. 161–9

Watson, J.A.L., Lendon, C. & Low, B.S. 1973, 'Termites in Mulga lands', *Tropical Grasslands*, vol. 7, no. 1, pp.121–6

Wells, A.T., Forman, D.J., Ranford, L.C. & Cook, P.J. 1970, 'Geology of the Amadeus Basin, central Australia', *BMR Bulletin*, vol. 100, Bureau Mineral Resources

White, M.E. 1986, *The Greening of Gondwana*, Reed Books, Sydney

Whitley, Gilbert P. 1972, 'Rains of fishes', *Australian Natural History*, vol. 17, no. 5, pp. 154–9

Williams, W.D. 1980, *Australian freshwater life*, Macmillan, Melbourne

Williams, O.B. & Calaby, J.H. 1985, 'The hot deserts of Australia' in Evenari, M. et al (eds) *Hot deserts and arid shrublands*, Elsevier, Amsterdam

Williams, W.D. 1990, 'Salt Lakes: The Limnology of Lake Eyre', in Tyler, M.J., Twidale, C.R., Davies, M. & Wells, C.B., *Natural History of the North-east Deserts*, Royal Soc. of S.A. Inc., Adelaide

York-Main, Barbara 1956, 'Taxonomy and biology of the genus *Isometroides* Keyserling (Scorpionida)', *Aust. J. Zoo*. vol. 4, no. 2, pp. 158–64

INDEX

Bold type denotes illustration or map